D0848577

Dear County Agent Guy

Calf Pulling,
Husband Training,
and
Other Dispatches
from
the Heart
of the Midwest

JERRY NELSON

WORKMAN PUBLISHING • NEW YORK

Copyright © 2016 by Jerry Nelson

The essays in this book originally appeared in a slightly different form in the column "Dear County Agent Guy," which was featured in newspapers across the Midwest and on websites. Some of the essays appeared first elsewhere: "That Old House," "Don Quixote, Tax Reformer," and "My Shameful Affair with the Farm Program" were published in *Successful Farming*. "The Ghosts of Horses Past" and "Uncle Wilmer" were published in *Progressive Farmer*. "Farm Corporate Jargon" was published in *Farm Journal*.

Library of Congress Cataloging-in-Publication Data is available.

ISBN 978-0-7611-8727-1

Design by Lisa Hollander

Photo Credits: Fotolia: Oleksandr Bebich vi, 71; maksymomicz: xiv; Getty Images: pshaun/E+ 132; Dover Publications, Inc.: v, vii, 1, 133.

Workman books are available at special discounts when purchased in bulk for premiums and sales promotions as well as for fund-raising or educational use. Special editions or book excerpts can also be created to specification. For details, contact the Special Sales Director at the address below, or send an email to specialmarkets@workman.com.

Workman Publishing Co., Inc.
225 Varick Street
New York, NY 10014-4381
workman.com

WORKMAN is a registered trademark of Workman Publishing Co., Inc.

Printed in the United States of America
First printing April 2016

10 9 8 7 6 5 4 3 2 1

To my wife, Julie,

the mother of our two sons

and

my biggest fan

Contents

~~~~~~~~~~

PART 1

Living the Country Life,
or Why "Let's Get Plowed!"
Doesn't Mean the Same Thing
to City Folks as It Does
to Farmers

1

PART 2

How to Raise Farm-Fresh Kids in Twenty-Five Years or Less!

71

PART 3

Never Kick a Fresh Cow Pie:

Lessons Learned from a Lifetime of Dairy Farming

133

Introduction

~~~~~~~~~~~~~~~~~~~~~~~~

I am the great-grandson of sturdy Norwegian pioneers who homesteaded on the open prairies of eastern South Dakota. Growing up on our 160-acre dairy, my sole ambition in life was to become a farmer like my dad and his father and his father before him. No one knows how far back my family's farming roots might reach, but it's entirely possible that my ancestors used mastodons to pull their plows.

Shortly after graduating from high school, I was able to rent a small dairy farm and begin my own dairy operation. For the next few years, I led the carefree life of a Norwegian bachelor farmer. Having grown up among five sisters, I had underestimated the pleasures of living without feminine rule-makers. Bathing became an option that was exercised perhaps once a month. A guy could scratch whatever he wanted whenever he wanted without fear of being called "gross" or being told, "What is *wrong* with you? Take that outside!"

But life without female companionship has its downsides, one of which is deep loneliness. I met my future wife and somehow managed to convince her to marry me. By the time we were thirty, we had two young sons, had purchased my great-grandparents' farm, and had built a new dairy facility in partnership with my parents. We were half a million dollars in debt and thoroughly convinced that we were living the American farming dream.

"Busy" does not begin to describe a dairy farmer's work schedule. City folks talk about putting in a nine to five; a dairyman's day commonly stretches from five to nine and beyond. Between chores and the kids and the cows and field work, there's barely time for bathroom breaks. This is why many farmers still drive tractors that have open platforms. A busy farmer also has a lot of hard-won knowledge regarding windage.

One blazing July morning, after we had been married for seven years, I descended into an underground manure pit to unplug a stubborn manure pump. I started feeling woozy and had begun to climb out of the pit when everything faded to black. Hydrogen sulfide, one of the most toxic substances known to man, had claimed another victim.

Or so it seemed. My father discovered me floating faceup in the manure a short time later. Emergency crews were summoned; hero stuff took place. After six weeks in the hospital—including most of a month in intensive care—I was able to walk out of the medical facility unassisted. Nurses cried, doctors were astounded. The only lingering effect, aside from the chest tube and tracheotomy scars, is some minor damage to the vision center of my brain. My peripheral vision isn't what it used to be, but the way I see it, that's a small price to pay for the privilege of being able to walk around atop the earth instead of reclining beneath it.

In the aftermath of my farm accident, many said, "God must have a plan for you!" Well, fine. But what, exactly, might that be? What would warrant such a cosmic tap on the shoulder?

The only thing that I know for certain is that the experience left me with a heightened appreciation for life. Every day on the topside of the dirt has been a bonus, a freebie. It's as if I totally blew off every single class yet still managed to graduate with an A.

Soon after returning home from the hospital, I went back to work on our dairy farm. Life went on; our sons continued to grow; the little blue ball kept on spinning. And through it all—a barn fire, being attacked by a bull, bankruptcy, my father's sudden passing—I tried to maintain a sense of perspective. I am not even supposed to be at the table; who am I to complain about the snooty waiter and the lousy service?

The summer of 1996 was wetter than a duck's butt. My fields were so soaked for so long that cattails had begun to grow where there should have been rows of corn. One of the main problems with wet weather is that there's nothing a guy can *do* about it. Except complain. Because there is so much that's beyond a farmer's control, many have become expert complainers.

Frustrated by the weather situation, I penned a letter to Mel Kloster, a friend and a Brookings County Extension Agent. A county extension agent is a purveyor of information, a conduit of knowledge between farmers and the state's land-grant universities. I asked Mel if he knew of a cheap, effective herbicide that would eliminate the cattails growing in my cornfield. And while he was at it, maybe he could tell me how to get rid of all those pesky waterfowl and Jet Skis that were out in my corn. Certainly they were tearing up some of the corn's leaves! (I was just guessing about the leaf damage part, what with everything being underwater.)

Mel said that he had enjoyed my epistle and encouraged me to publish it somewhere. My first reaction was "Why the heck would I want to do that?" My second one was "Why the heck not? I should be pushing daisies at this very moment; who cares what anyone thinks?" So it was pretty much on a lark that my column "Dear County Agent Guy" was born and I began my voyage across the terra incognita of journalism.

My formal training as a writer began and ended with a high school class called Creative Writing. One day our teacher, Mr. Brown, announced that we had thirty minutes to craft an original essay. I dashed off a fantastical tale about a nerdy farm boy meeting and dancing with a stunning young woman. After reading it, Mr. Brown loudly praised my story, causing me vast amounts of embarrassment. I was also secretly gratified. Despite what Mr. Brown had said about my writing abilities, I never dreamed that committing my words to paper would be of any value. After all, few of the farmers I grew up with would have viewed writing as real work. I have since come to see that writing can be just as much work as plowing a field and that a harvest of well-tended paragraphs can be every bit as gratifying as filling your bin with grain.

I have been blessed with numerous opportunities to offer up my thoughts for the world's scrutiny. And while my wife and I no longer dairy farm, we continue to harbor fond memories of the dairy farming business. Well, some of them are fond. Let's just put it this way: Nothing good ever comes with a wet tail to the face.

Part 1

Living the
Country Life,

or

Why "Let's Get Plowed!"
Doesn't Mean the Same Thing
to City Folks as It Does
to Farmers

A Norwegian Bachelor Dairy Farmer Finds a Wife!

~~~~~~~~~~~~~~~~~~~~~~~~~

I have, over the years, received numerous cards and comments directed to my wife. These cards and comments all have one particular message in common: "You poor dear! How on earth do you manage to put up with that mean old husband of yours? He picks on you in his column all the time!"

One little old lady reader, upon meeting my wife and me at a farm show, blurted out to my missus, "You poor thing! I ought to send you a sympathy card!"

Hold on just a minute here! I admit that I pick on my wife some, but it's all in good fun. Plus, she's of resilient Teutonic stock and can handle just about anything.

But why do I do it if it causes so much trouble, you might ask. The answer to that is complicated, but boils down to the fact that she deserves it. Sorta. Kinda. From a certain point of view. Here is an example of what I'm talking about:

My wife and I had been going steady for a short while when I decided to drop in on her unannounced one evening. After I finished milking and cleaned myself up a bit, I drove into town and found the little trailer house she was living in at the time. Imagine my surprise when I discovered a strange car parked outside.

I instantly assumed the worst, that she was in there with some other man and was on an intimate basis with him. This was due mostly to the size of her trailer house, which was so small that you couldn't help but be intimate whenever there was more than one person inside.

I asked myself a momentous question. "Self," I said, "should I simply turn around and go home and never see her again? Or, should I go in and fight whoever is in there—it could very well be Jackie Chan, after all—for my girl?"

I sucked it up and did the manly thing: I got out of my pickup, walked up to her door, and knocked. I will admit that I left the pickup running, though. If you've ever seen any of his movies, you would know that Jackie can be pretty feisty when it comes to fighting.

My girlfriend answered the door and seemed quite happy to see me (in retrospect, I should have suspected something). After greeting me, she said those three words that can wither the manhood of even the manliest man. "Meet my dad," she said.

But that wasn't the worst of it. As she introduced us, she informed me that her father was an IRS agent. So it was that I spent an uncomfortable evening squirming and chatting and trying to be friendly and feeling profoundly guilty about a couple of different areas of my life.

Another example happened about a year later, on our wedding night. We had done the whole church-wedding thing and the whole wedding-dance-afterward thing. It was two o'clock in the morning when we got back to our little farm; the plan was to toss on a change of clothes, grab our suitcases, and take off on our honeymoon trip.

As we drove down the driveway, we saw numerous blue-green dots of light moving about in the dark distance. "What's that?" asked my brand-new bride. "Are those fence post reflectors?"

I felt a twinge deep in my gut. "If only!" I replied. "But fence posts don't move around like that."

It was cows. My cows. They must have gotten together and decided it would be a hoot to bust through the fence

4

and thus oblige me to chase them around in the dark on our wedding night.

I parked the car and strategized. "Okay, here's the plan," I said. "You go down by the barn and hold the gate open. I'll chase the cows toward you, and if any try to get past, you just flap your wedding dress and yell. Watch your step, though; there's gonna be fresh cow flop lying about."

My wife looked down at her white gown, then fixed me with a steely stare. "If you think I'm going to chase cows in the dark in my wedding dress, you're more Norwegian than I thought!" She then got out of the car and stomped into the house.

Six hours. We had been married a whole six hours and already she was flouting the "obey" part of our wedding vows.

And that's why I pick on her from time to time: because she's been one up on me for the past thirty-some years. I figure I might begin to pull even in, oh, another thirty years or so.

A recent random conversation with a dairy farmer gobsmacked me with some ancient history.

This particular dairy farmer has a particular surname, one that prompted a primeval memory to slowly rise from the murky depths of my gray matter, causing me to casually ask if he knew of a woman named Rosie who shares his surname.

"Of course I do," he replied. "She's my sister."

I was stunned; you could have knocked me over with the flick of a cow's tail. "Your sister," I managed to mumble, "is responsible for me being married."

Back when I was a young and struggling dairy farmer, I was actually struggling on two fronts: First and foremost was making a go of it as a dairy farmer. Following at a very close second was finding female companionship.

One fall evening I was walking down the street of a nearby town when I happened to bump into Rosie. She and I had a bit of a history, having hung out some at the local roller rink. We had also gone out on exactly one date. Every time I asked her out on a second date, she said she couldn't go due to an urgent hair-washing emergency. Just my luck.

"Rosie!" I exclaimed that night when I met her on the street. "I haven't seen you in a coon's age! What say we duck into this fine establishment and I'll buy you a beverage?" Rosie allowed that this sounded like a swell idea.

We sat at an empty table and I ordered refreshments. Rosie then got up, saying that she needed to visit the ladies' room. I told her that I would keep her chair warm.

I have not seen her since. It's been well over thirty years now and I'm beginning to think that she isn't coming back. Actually, I came to that conclusion within about half an hour. I later noticed that the ladies' room was located near the bar's rear exit.

My buddy Steve happened to be in the joint that evening, so he and I commiserated about the ineffable enigma that is the female species. As we talked, we began to construct a pyramid of empty beer cans, most of which were requisitioned from nearby tables.

A waitress stopped at our table and I thought, "Welp, there goes our pyramid." But instead she left us several more empties, enabling us to build a veritable Tower of Babel of beer cans.

I thought this was pretty nice of her, which prompted me to stop and chat with her on my way out. That encounter went a bit better than the one with Rosie, as the nice waitress and I have now been married for more than three decades and have two grown sons.

I never did figure out why Rosie gave me the slip that night. Perhaps she decided, at that exact moment and for no good reason, that she really didn't like men all that much. It pained me to think that I may have been the last male contact for such a nice young lady, who, at that very moment, had decided to lead a life of celibacy.

The chance encounter with Rosie's brother caused me to wonder what might have been. After all, her ditching me in that bar wouldn't have been a deal breaker, as far as I was concerned. Beggars and choosers and all that.

I observed that Rosie's brother's teenaged kids are tall and good-looking and muscular, with straight white teeth. But so are the two sons who were born to my wife and me. I then made an inquiry that brought forth a crucial fact: Rosie's family is of Irish extraction!

The mind reels with the possible complications that might have arisen from a union between a woman who is Irish and a guy who is Norwegian. For instance, can you imagine the nuclear odors that would radiate from a household where both lutefisk AND corned beef and cabbage are consumed? And what's to keep a certain person from desecrating a perfectly good hunk of lefse—the Norwegian version of a tortilla—by dredging it through her mulligan stew?

Ruminating upon these things, I deemed myself extremely lucky. How fortunate that Rosie ditched me that night, causing me to meet and marry my wife! How providential that I thus found someone who so sweetly tolerates my many foibles—even my "secret" pickled herring habit!

As I parted company with her brother, I casually asked, "So, what's Rosie up to nowadays? I imagine she's living in a convent or something."

He looked at me as if I had just taken a swig of kerosene. "Rosie is married and has four kids," he said. "Whatever makes you think that she's living in a convent?"

The Throwback

~~~~~~~~~~~~~~~~~~~~~~~~~~~~

The tale goes something like this: A pair of Norwegian homesteader brothers built themselves a nice granary, two stories tall, with a sturdy stairway. Come oat harvest they bagged their grain at the threshing machine, hauled it to the granary, then carried the bags upstairs so they could dump the oats into the bin below.

Being Norwegian, it never occurred to them to simply walk into the bin and dump the grain on the floor.

Some will slap their foreheads and say, "Uffda! That's the sort of story that gives us Norskies a bad name!" I, on the other hand, think that they were merely throwbacks. And I should know throwbacks, because I am one.

For instance, some perfectly good bromegrass grows in the ditches that run alongside our township gravel road. This grass will be useful next winter when our half dozen Jersey steers want something to munch on other than the icicles hanging from the barn roof.

The most efficient way to harvest this hay would be to hire someone with a modern hay cutter thingy. Taking this path would require no more effort than changing channels on the TV.

Being a throwback, I chose a different path. When the hay is ready to cut, I fire up my venerable John Deere A tractor, which was manufactured at the close of the Second World War. I hitch the A to my No. 5 sickle mower, a machine that bears a patent date of 1936.

Then I mow our farmyards and our road ditches, jostling over the innumerable humps and bumps. The A has Armstrong power steering, so every little pebble is telegraphed through the steering wheel to your hands, up your arms and shoulders, and into your brain. This is why we throwbacks tend to have a subpar IQ.

The sickle inevitably bites into something that doesn't agree with it and it loses one of its triangular sections. Cutting hay with a busted sickle section is like eating sweet corn with a missing front tooth: neither pretty nor efficient.

The sickle must be removed so that a new section can be riveted in place. The art of riveting has been around for several thousand years, which makes it a favored activity for us throwbacks.

Flattening an iron rivet with a ball-peen hammer is extremely satisfying—as long as your fingers don't become part of the flattening process. Percussive maintenance is a tremendous outlet for frustrations.

"Here's for being forced to change that tire in the rain!" you mutter as the hammer mashes the rivet. "And here's for the signal dropping out during that important call!" you grunt as the rivet squashes down even farther. A bad day frequently leads to over-flattened rivets.

One must often adopt a Zen attitude while working with old equipment. When the mower broke and needed a new part, I didn't become irate upon learning that they didn't have the part at the dealership. I instead celebrated the fact that the part actually existed and would arrive in a few days. The mower may be slow, but the grass is patient. And becoming impatient with the parts guy wouldn't help.

As I mowed, the neighbor across the road was harvesting his alfalfa. Comparing his rig to mine would be like comparing the space shuttle to a sparrow. He was buzzing along at speeds normally associated with NASCAR.

I could have hired him and he could have done the job in two minutes instead of my two hours. But what would I learn from that, other than how to write a check? I would like to avoid practicing that particular skill.

Once the hay has cured, it must be raked. Again with the bumps and the jostling. Paint could be shaken.

After the hay is raked, I call my neighbor Ziggy, who rolls it up into big round bales. Putting the hay into small square bales would be more retro. But I'm somewhat of a lazy throwback.

Moving the big round bales involves hitching the A to a two-wheeled cart that was constructed from old oil pipe by a local machinist. The bale is lifted via a hand-cranked winch—a word my Norwegian grandfather pronounced as "wench." I imagine a really strong wench could also perform this task.

Squirreling the hay into storage evokes a deep sense of satisfaction.

"Look at that!" I exclaimed to my wife, indicating the hay. "Those bales mean we won't starve!"

"What, are we going to eat hay soup?"

"No, the steers will eat the hay and we'll eat the steers. We have it made."

"You are such a throwback," she replied with a shake of her head.

It's not often that I receive such high praise. ●

Silo Time

~~~~~~~~~~~~~~~~~~~~~~~~~~

Q uestions commonly heard at our house during a particularly long winter might include "Will this winter never end?" and "How far is Key Largo?" and "When's the next flight?" and "Tell me again: WHY do we live here?"

Regarding the "why" part, I patiently and repeatedly explain to my wife that we endure these winters because it keeps out the riffraff. A person has to want to live here. Plus, beastly winter weather filters out those silly folks who believe that man can live on salad. Here in the frozen North, it's survival of the fattest.

Back when I was a kid, every single winter—without exception—was like an extended stay at the North Pole. I recall chest-deep snow, howling winds, and temperatures normally associated with Siberia. And that was just in my bedroom. Conditions were much harsher outside.

And outside is where we spent most of our time. This was an era when conventional parenting wisdom held that children were invariably better off out of doors. Are the kids fighting? Put them outside. Is your child convulsed by the croup? Send him outside. Does your teenager have what appears to be a terminal case of zits? Get him outside!

And since they're outside, they may as well be doing something useful, such as chores. At least that's how it worked at our place. Everyone at our dairy farm—all eight kids and both our parents—participated in the family activity called chores. At chore time, we swung into action like a well-oiled machine, albeit one that revolved around a fleet of rusted and dented five-gallon buckets.

At about the age of ten, the task of muscling the silage out of the silo fell to me.

Throwing out silage involved shoveling silage out of our silo, a concrete cylinder that's fifty feet tall and sixteen feet across. It was the first measurement that made this chore such a chore.

Unlike with most jobs, a person who unloads a silo begins at the top. The unloader (me) had to first climb up the claustrophobia-inducing tunnel-like galvanized chute that ran the length of the silo. The most crucial component for shoveling silage was the silage fork, a hand tool that is essentially an oversized garden fork. Scooping silage with the more common four-tined pitchfork would be akin to trying to empty the ocean with a teacup.

Each September, we used a roaring, tractor-powered blower to fill the silo with chopped corn. After the fodder had fermented for a month, I would ascend the chute with the end of a long baling twine in one hand. The other end of the twine was tied to the silage fork, which I pulled up the chute once I reached the top of the silo. The first trip up the silo always had an element of surprise because you never knew how much the silage had settled. The cylinder of silage could have shrunk a foot, or maybe it had dropped ten feet. Opening the silo was like unwrapping a birthday present.

Each evening I would clamber to the top of the silo and crawl into it through a man-sized rectangular doorway. Anyone who has ever ascended a silo chute knows that the wind is nearly always blowing up it. A silo chute is, in reality, a vertical wind tunnel. The turbulent air causes particles of silage to swirl around the climber; itch-inducing silage finds every possible crack and crevice.

When I finally reached the top of the silo, I would often need to pause to spit out bits of silage. I cannot understand why cows like the stuff; silage tastes like pickled bad breath.

Throwing all my weight into it, I would jam the silage fork into the silage, which gravity had packed into a substance that closely resembled concrete. Back muscles moaning in protest, I would pry up a slab of the fermented fodder and heave it through the doorway and into the chute. The wind sometimes worked so powerfully against me that it seemed as though the silage was blowing back up faster than I could throw it down.

I might pause for a breather and mull things over, and it would occur to me that I was responsible for the lives of ten people: If I didn't throw out silage, our dairy cows wouldn't give milk and we might all starve.

As the Arctic winter deepened, silage would begin to freeze to the concrete staves. I did my best to keep the walls clean, but a silage fork is fairly feeble in the face of ferocious frost.

A tiling spade then had to be hauled up into the silo. This spade was used to hack a slot that allowed the doors to be removed. As the silage froze inward, the diameter of the silo narrowed. I felt like a prisoner who was digging an ever-shrinking dungeon.

Winter thaws were welcome but also hazardous, as my towering silage walls would begin to weaken and tumble. Much of the fallen silage was still permafrost, so it had to be hacked into smaller pieces before being sent down the chute.

I would sometimes watch the goings-on at the bottom of the tunnel and try to eavesdrop on the murmured conversations. Hands would scrabble into view as my siblings packed silage into five-gallon buckets. It looked as if gloved spiders were stealing the fruits of my labors.

At such times I might drop a chunk of frozen silage down the chute. The chunk was usually caught by the updraft, causing it to tumble and thunder against the steel tunnel. But sometimes the projectile would fall straight and silent. Its sudden impact would elicit startled mutters down below.

This was all I had for entertainment during my silage-throwing chores. That, and perhaps daydreaming about such places as Key Largo, where, I assumed, silage never freezes to the walls of their silos.

That Old House

I have two houses on my little farm. It's not that I am among the wealthy elite; quite the opposite. It's more that I am a victim of my frugal Norwegian heritage.

My great-grandfather Charlie Sween emigrated from Norway and homesteaded this 120 acres of eastern South Dakota more than 120 years ago. After living in a crude sod shanty for some time, he was able to afford a real house, a house made of wood, with a brick chimney and glass windows. A place to call home.

He spent the next fifty years living in that house, eking out a living and raising a family. One of his daughters, Elida, would eventually become the bride of my grandfather, Erwin. In the mid-1930s, Erwin bought this little farm from his father-in-law. Erwin and Elida's union produced seven children, one of whom was my father.

In 1963, at age sixty-five, Erwin decided to tackle a crazy project: He wanted to build a new house. Family and friends ridiculed him for taking on such an endeavor. "Move to town," they told him. "An old farmer like you doesn't need a new house." Instead of heeding their advice, he hired a crew of carpenters.

He wasn't quite sure what to do with the old house. It seemed a waste to demolish it, but something had to give, since it occupied the space where the new house would sit. In a moment of inspired frugality, he hired a bulldozer to push the old house far out into our grove of trees. And there that old house sits to this very day.

Erwin and Elida passed away, and we purchased the farm from their estate. My wife and I raised our sons on this place and have lived here for more than thirty years. Our farm is located just across the section—two miles by road—from the farmstead where I grew up and where my parents lived and farmed.

When we first moved in, my wife took one look at the derelict house in the grove and declared it a hazard and an eyesore. I agreed and planned on a colossal bonfire once conditions were right. One autumn day, I thought that perhaps the time had come for the old house to finally have its Viking funeral. But I deemed it prudent to check out the old shack first, just in case something of value was left behind.

Our two young sons tagged along, and we waded through the tall grass in the small meadow where the old house sat. Time had taken its toll. The front porch had collapsed in on itself, most of the windows were gone, and the siding was rotting and falling off. We entered through an open window and got the olfactory impression that skunks had resided beneath the floorboards.

I felt as though I had stumbled onto a forgotten time capsule. Here lay the sundry detritus of my grandparents' lives: A broken chair. Some old clothes in a gunny sack. A thermometer from a grain elevator. But the thing that drew my eye was a cardboard box stuffed with papers.

I dug through the contents of the box and was instantly transported back in time. Numerous items begged my attention. A tax return from 1957. An aunt's first grade report card. Canceled checks from June of 1962. Greeting cards from old friends and relatives, now all dead and gone. An uncle's third grade spelling book.

I spent most of a pleasant hour going through the contents of that cardboard box. All the while, I had to answer a stream of questions from my sons about the old house and who had lived in it. They were amazed that nine people had once occupied that tiny structure, and that they did so without running water or electricity. I related to them how my father remembered that on cold winter mornings, a pail of water would be iced over even though it sat right next to the cookstove. And they shivered when I told them that in those days the cookstove was their only source of heat.

So it was that the old house was spared the torch. A few times a year I would feel the old house calling, and my sons and I would venture in once again. We felt like archaeologists investigating the ruins of a lost civilization. We were never disappointed; each time, we would retrieve new treasures: old farm magazines, a 1949 calendar, antiquated school textbooks. And one special time, I uncovered an old Bible, written in Norwegian and set in Gothic type, that had "Charlie Sween" handwritten inside its cover. Such a find could have no price.

The years passed, and our visits to the old house became less and less frequent. We were too busy, it seemed, and the old house once again enjoyed the lonely solitude of the meadow in our grove. As we hurried through our lives, I might catch a glimpse of the old house through the trees and wonder: How did they manage? How did they survive the dust storms and the floods and the blizzards and the Great Depression? They must have been made of sterner stuff than me.

I remember how, as a child, I would struggle to walk in my father's footprints. Even then, I could imagine no nobler calling than being a farmer just like Dad. As a man, I was afforded the opportunity to enter into a partnership with my father on his dairy operation. Daily, we toiled side by side through drought and deluge, through good times and bad. And, despite it all, we somehow managed to survive.

Then, one April morning, my father was felled by a massive heart attack, at age sixty-eight. The entire family was shocked by his untimely passing, none more than me. Dad was not just my father, but also my coworker, my business partner, my trusted adviser. And now he was gone, so suddenly gone.

I found myself thrust into a new position in life. Mine was now the last word when it came to farming decisions, and I had abruptly become the eldest male in our family. I wasn't exactly comfortable with either.

There are some things in life that are unexplainable. Why I ventured out to that old house on that day shortly after my father's funeral is still beyond me. It was as though it were calling; even the trees seemed to whisper an invitation to come, to visit, to tarry awhile.

As I stood once again on that ancient linoleum, my eye was drawn to a jumble of papers on the floor. An envelope, yellowed with age, lay on top. A blue stamp on the envelope read "Cleared By Military Censors."

How could have I missed this artifact? My father had served aboard the USS *Washington* during World War II and had written home whenever he could. My grandmother had saved all his letters.

I carefully removed the letter from its envelope. It was dated September 1944. My father would have been somewhere in the South Pacific at that time and all of eighteen years old. I studied the familiar scrawl. Dad wondered how the oat harvest had been and how his uncle's new team of horses was working out. He supposed that his youngest brother was starting first grade and imagined that he was becoming quite the little man. He asked his mother to greet everyone and said that he missed them all.

It wasn't hard to read between the lines. Here was a homesick young man, a kid, really, who had spent his entire life dwelling upon a sea of prairie grass. Now he was on a different kind of sea, an ocean that was being roiled by the thunder and the lightning of a world at war. Until then, his life had been focused on nurturing and caring for life. Now, he had been schooled in the fine art of mayhem and had become a cog in the wheel of a world-class killing machine.

At the bottom of the page, in underlined print, my father had passed on one last message. Tears burned my eyes as I read those words he had so carefully emphasized: "All is well here. Please don't worry. I am doing fine."

As I left the old house that day, I took one last glance back at it over my shoulder. I don't care what anyone thinks, I decided. That old house gets to stay there until it rots into the earth.

The Modern
Marvel

~~~~~~~~~~~~~~~~~~~~~~~~~~~~~~

T he other day I was watching the Discovery
Channel when a program came on that purported
to be all about mankind's technological progress.
They talked about stuff like space travel, the polio vaccine,
and the discovery of the element "spandex."

Shows what they know. They didn't even mention the
calf puller.

For those who may be unfamiliar with this device,
here's a brief description: Remember the medieval torture apparatus they called "the rack"? It was used chiefly
to extract "confessions" from the "guilty." This was accom-
plished by tying your arms to a large, winch-like machine.
Your legs were chained to something massive and
immovable, like the footings of the dungeon. If you were
a particularly tough customer, more extreme measures
would be taken, such as locking you in a room with Justin
Bieber.

Your cheerful host would then crank the winch until

A) you could slam-dunk without even jumping or B) you would shout, "I admit it! I kidnapped the Lindbergh baby! And I left the toilet seat up! Just don't make me listen to 'Mandy' by Barry Manilow again!"

Now imagine such a contraption shrunk down and modified so that it can be used to help a cow who is having trouble giving birth.

Calf pullers hadn't yet been invented when I was a kid. If a cow's delivery was perceived as being overly difficult, Dad would summon us eight kids to the barn.

Dad would tie a baling twine to the calf's front legs (we were so poor, we couldn't afford to buy a real rope) and tell us kids to grab on. Dad would situate himself next to the cow so that he could observe the proceedings and give instructions.

We would haul away on the twine until our fingers felt like they were breaking and our arms threatened to pop out of their sockets. Calling a veterinarian to help was deemed an outrageous extravagance, on par with sprinkling caviar on your oatmeal.

One day, when my wife and I were still pretty much newlyweds, I noticed that one of our heifers was lying on her side out in the cattle yard, striving mightily to deliver a calf. I could see that the heifer had little to no chance of succeeding without assistance. It was as if a Mack truck were trying to squeeze through the kitchen door.

I summoned my "town girl" bride and dug through the twine pile until I found a good, strong baling twine. I had just snuck the slipknot over the calf's front legs when the heifer glanced back and suddenly noticed that a pair of creepy humans were messing around with her nether regions.

The bovine sprang to her feet and tore across the mucky cattle yard at a brisk gallop. She seemed not to notice that my wife and I were being towed along behind her. It was kind of fun, actually.

Sort of like water skiing, only in mud.

The heifer would stop to push whenever she was overcome by a contraction. My wife and I pulled so hard that our fingers turned purple and our arms lengthened by several inches.

Did I mention that my wife was eight months pregnant at the time? I figured, what the heck, the more ballast, the better.

The heifer dragged us alongside the fence, no doubt hoping to shred her human tormentors on the barbed wire. When she stopped for a contraction, I whipped the twine around a wooden post and took a quick wrap. When the contraction was over, the heifer tried to run off and the calf popped out like a cork from a champagne bottle. The heifer then galloped away, leaving us to deal with her sputtering newborn.

My wife looked down at her manure-spattered clothes. "Are you sure this is how it's done?" she asked.

"Nope, we got lucky this time. The next one will likely be a lot tougher!"

"Well, I'm not going to do this ever again!" she snapped as she slogged out of the cattle yard. She was so upset, I figured it wouldn't be a good time to point out that she had lost a shoe.

But she was right. The next day, a sparkling new calf puller mysteriously appeared in the barn. A debit in the checking account for approximately the price of a calf puller appeared later that week.

I guess that's the price of progress. 🍂

Adventures
in Cow Herding

~~~~~~~~~~~~~~~~~~

Avast assortment of slogans pass for wisdom on bumper stickers, but I like one I saw many moons ago that declared, "Cows don't give milk. It's taken from them by force."

That sentiment could be interpreted a couple of different ways. Being a dairy farm kid, my first thought was "Right on. Dairy farming is a tough job and it's about time we got some recognition."

I was so moved, I considered writing a letter to my congresspersons stating that dairy farmers deserved to be honored. But then someone informed me that this is why we have June Dairy Month.

When I was a kid, the arrival of June meant many good things. It meant no more school, no more morning bus to catch, the suspension of scheduled bathing, and generally reverting to an untamed state. But above all, June meant the beginning of the grazing season.

We had a pasture, but also took advantage of the free fodder growing in our road ditches.

The ditches that bracket our roads are a gift from civil engineers. Roads—even the humblest gravel avenue— were all brought into being via a long process of scraping and packing and grading the soil. A common practice for a township thoroughfare was to construct a "two-rod road." This is a road that measures two rods—thirty-three feet— from each side of dead center.

The ditches that keep our township gravel roads drained are somewhat of a no-man's land. The landowner pays taxes on all the ground up to the center of the road, but because of erosion concerns, isn't supposed to farm past the imaginary two-rod line. However, haying or grazing the ditches is perfectly acceptable.

Grazing cattle in the road ditches was actually pos- sible when I was a kid due to the fact that barbwire fences still marked every two-rod line. Back then, it was unimag- inable that farmland would ever become so valuable that fences would be ripped out so that every possible square inch could be tilled.

Herding our Holsteins in the ditches took a bit of planning, along with a crack cow-handling team. This team often consisted of my sister Di and me.

To the layman, our job may have seemed simple, con- sisting mainly of heading out ahead of the herd as they munched their way southward from our farmstead. When the cows reached the end of our gravel road, it was up to us to turn the bossies around and encourage them to head back toward home.

We had tools to help us with this very vital job, the main ones being sticks that we would wave threateningly at the cows. These were backed up by pebbles gathered from the edge of the gravel road. If our shouting and the waving of our sticks didn't convince our cows that they needed to turn back, we would reinforce our message by winging pebbles at them.

Herding was usually a fairly tame experience. Di and I had ample time to talk about all manner of things as we kept a weather eye on the cows and enlarged our pebble arsenal.

The cows would at first frisk around, relishing their newfound freedom. But it wasn't long before they became serious about the task of gobbling grass. As the cows gulped down the luscious greenery, Di explained how cattle have the ability to eat now and chew later. As a teenager, I attempted to duplicate this feat numerous times but without success.

One June, when I was maybe seven, Di and I were making our way south from the farmstead. King, our

German shepherd, had volunteered to tag along; owing to the "shepherd" part of his pedigree, we assumed he would prove an able assistant.

As we passed a culvert, King began to snoop obsessively at one end of the huge tube. A horrible cacophony of snarls, squeals, and growls erupted from the ditch. King had engaged in battle with a giant raccoon.

A blur of fur rolled up onto the gravel road. It was difficult to see who was winning; the raccoon was nearly as big as King and was obviously an experienced scrapper. Di and I could do nothing but watch from what we thought was a safe distance, gathering large pebbles and hoping that the melee would resolve in King's favor; we had no idea how an angry and wounded coon might react to the sight of a pair of kids armed with sticks and fistfuls of pebbles.

News of the battle thundered across the prairie on an expanding shock wave of roars and yelps. The cows halted their grazing and froze in midchew. They stared with bovine fascination, clumps of slobbery grass dangling from their mouths.

The hurricane of caterwauling boiled across the steaming gravel road until King at last gained the upper hand. Seeing an opening, he seized the colossal raccoon by its throat and pinned it to the ground. The masked varmint was soon reduced to an inert bag of mangy fur and broken bones. Our cows, sensing that the show was over, calmly resumed grazing.

Such a thing never happened again, so I guess the word must have gone out in the raccoon community that King was indeed *Canis rex*.

Remarkable and scary as that incident was, it wasn't without an upside. Because it proved to me that producing milk can sometimes involve a great deal of noise, along with some very high drama. ●

Domestic Lessons

~~~~~~~~~~~~~~~~~~~~

My wife says that when we first met one of the things she found attractive about me was that I had been "pre-trained." Specifically, she found it appealing that I had grown up in a family of five sisters and had thus been exposed to the lofty standards of behavior that females expect, such as no spitting in the house and closing the bathroom door before you settle in to make a boom-boom.

In addition to such things, we boys (I have two brothers) were also taught how to cook and do laundry. If you were a male growing up on our dairy farm, you had to possess a Swiss Army knife–like skill set, capable of sorting cattle in the morning, clothes in the afternoon.

I'm not complaining. On the contrary, learning how to perform domestic chores came in handy when, for a few years, I led the life of a Norwegian bachelor farmer. It was useful to know which household tasks—such as extinguishing a grease fire—required immediate attention and which jobs could be put off until later. For instance, who, exactly, cares about the petrified spaghetti that has glued itself to the couch cushion? It's not as if the president is going to drop by!

One of the earliest domestic lessons I learned was how to whip up enough pancakes to feed ten people.

It somehow became a tradition at our house to have a Sunday evening supper of pancakes and bacon. Perhaps it was a subliminal echo of Sunday morning church services with its Communion wafers and syrupy wine.

When Sunday evening's cow-milking operations hit the halfway mark, which meant there was less than an hour left before we were done, I would be sent to the house to construct a pancake supper. Our pancake recipe is a snap: two of everything except for the sugar, which is a third of a cup.

Getting the batter right is an art. Too runny and your pancakes will be thin as Bible leaves; too thick and your flapjacks will be tough and doughy and will have approximately the same density as lead.

We had a massive cast iron griddle that could accommodate six pancakes at a time. Gauging the proper griddle temperature is a marriage of magic and science. When a droplet of water (or spit; not that I would know anything about *that*) hisses and bounces across the griddle, it's ready.

As the pancakes piled up in quantities that were measured by the foot, I was also frying mounds of bacon. One
spring, Dad took a skinny old sow to the butcher shop and
she bestowed us with strips of bacon that were nearly a
yard long and lean as shoe leather. Just as tough, too.

The objective was for supper to be on the table by the
time milking was finished. We could thus be done eating by
the time *Bonanza* came on.

Bonanza was a popular television Western about a single dad who was struggling to raise a family under difficult
conditions. And by "difficult" I mean "on a replica of the Old
West frontier as envisioned by Hollywood set designers."

Bonanza detailed the travails of the red-blooded,
all-American Cartwright clan. Its rugged, steely-eyed
patriarch was portrayed by Lorne Greene, a Canadian actor
who was of Russian descent. The Cartwrights employed a
cook named Hop Sing, a Chinese gentleman who neither
hopped nor sang.

Our pancake supper would be narfed down swiftly
and our family would gather around the TV. We didn't want
to miss a millisecond of our time with the Cartwrights!

The cast of *Bonanza* were all guys, so there was a lot
of manly action, such as fistfights, gunslinging, and posse
chasing. Despite the distinct lack of feminine influences

in the Cartwright household, none of the guys ever sat on the couch and clipped their toenails with a hedge trimmer or excavated navel lint with barbecue tongs. They were a pretty genteel bunch.

A female love interest would occasionally pop up on the show, but the gals never stayed for long. Ben, the paterfamilias, had buried three wives; perhaps the ladies got wind of this and wisely decided that attempting to domesticate the Cartwright guys wasn't worth the risk.

Some years later, I was living the life of a Norwegian bachelor farmer when I acquired my own love interest. One Sunday evening, I invited the young woman out to my humble little dairy farm for a home-cooked supper.

I was in a tizzy regarding what to serve the young lady. But then my boyhood domestic training kicked in and I decided to whip up a batch of pancakes and fry a pound of bacon. When all was ready, I nervously set it before my lady friend. I was worried about what she would think of my feeble offering. Would she turn up her nose and proclaim that such a supper—and by extension, I—was lame?

"This is wonderful," cooed the woman who would soon become my wife as she poured syrup over a heap of flapjacks. "How did you know that we used to have this at our house every Sunday night?"

My New Address

~~~~~~~~~~~~~~~~~~~~

I've got a new address, don'tcha know.

Like many who live out in the country, we have recently been assigned one of those new-fangled "E-911" addresses. Up until a few years ago, our address was simply Rural Route 1, Box 38. This has been replaced by a set of impersonal numbers that are similar to those that are used for urban addresses. We now even have signposts at the intersections of our township roads that denote streets and avenues! What manner of urbanized hell will come next? A Starbucks at the end of my drive-way? Is Google Earth behind all of this?

In the name of modernization and efficiency, I am now burdened with yet another set of numbers to remember. As though I didn't have enough integers rattling around in my head with stuff like my Social Security and my driver's license, and then there's my wedding anniversary, which of course is ... um ... Excuse me for a moment. I seem to have this sudden urge to run out and buy my wife a romantic greeting card.

One of the problems I have with this new address

is the fact that it sounds so much less pastoral than the old "Rural Route" system. I often have trouble convincing city folks that I really do live out in the middle of the boondocks.

Another problem is the system they used for numbering the streets and avenues. Apparently, they started with a street called "First" somewhere around the equator and worked north from there. The net result for me is a street address that reads about like the value of pi when it's divided out a dozen places or so. It's enough to arouse concern about the so-called efficiency of this system.

Hysterical woman: We have an emergency! Send help!

Dispatcher: Keep calm, ma'am. Just give me your New and Improved address and we'll send someone right out.

Woman (near panicking): Um, let's see . . . It's 5280 31427th Street. No, wait. I think it's 31427 5280th Street. Yeah, that's it!

Dispatcher: Okay, so you're saying we should send someone out to 3180 52427th Street? Or was it 52480 3127th Street?

Compare that to the old system, where an emergency call might go like this:

Caller: Send help! This is Alma Klinkhammer and we live on the old Jensen place!

Dispatcher: Okay, Alma, stay calm. Now, which Jensen place is it? The "Tramp" Jensen place or the "Crazy" Jensen place?

Caller: I don't recall . . .

Dispatcher: Well, "Tramp" was the one who always had that three-legged dog following him around, and "Crazy" was the one who kept that big flock of sheep.

Caller: Oh. Send them out to Tramp's, then. Say, didn't those boys have a sister named Olga?

Dispatcher: Why yes, I believe they did. Isn't she the one who ran away with the Tattoed Man from the circus when she was sixteen?

Caller: Yeah, that's the one! And didn't their father have a glass eye? . . .

My wife, being a city girl, had her own problems with conquering the fine art of navigating out in the country. I soon discovered that telling her things like "turn at the Mannerud corner" didn't help her any.

Once, shortly after we married, I asked my wife if she could bring dinner out to me at a distant piece of land I was farming. She replied, "Sure, just tell me how to get there." I began instructing her as to how you go four miles south, then three west, when she interrupted me.

"Just tell me how many blocks it is and cut out that silly east or west crap. You think I'm a ship's captain or something? Just say where you go left or right like any normal person would."

My wife was eventually able to forge her own system for navigating around the countryside; if things had been left up to her, we would have street names like "Rock Pile on the Hill Avenue" and "Old Corn Picker in the Weeds Street." And the road we live on might be known as "Dead Skunk in the Ditch Boulevard."

But, alas, it appears as though we country folks, in the name of progress, are going to be stuck with these new and improved yet cold and impersonal addresses. Now I have another set of numbers to memorize, along with my zip code, checking account, blood type, wife's birthday, and . . . um . . . Just a minute here. I seem to have been seized by this sudden urge to call the florist and have something sent to my wife.

And you know, I think I'll tell them to deliver it out to our farm, which is otherwise known as the old Charlie Sween place. If the delivery driver doesn't happen to know where that is, just tell him that we live a mile east of Heppler's slough and half a mile north of the Mannerud farm. And should he become lost, simply send him the GPS coordinates of that new Starbucks at the end of my driveway. ●

Old Frank

~~~~~~~~~~~~~~~~~~~~~~~~~~

When my wife and I first started farming, we rented a half-section of land from an old guy named Frank. Frank was well into his eighties by then, and rumor had it that he was a millionaire.

Old Frank had a unique way of communicating with people. For instance, if he were to say, "I know a guy who should check his zipper," it probably meant that my fly was open. Or he might say, "I don't mean to tell you your business but . . ." and then he would go ahead and tell me my business.

His business advice was always augmented with an earthy, homespun yarn concerning a profitable exploit he had pulled off during his four score years of living. I found these parables enlightening and entertaining at first, but after hearing them a dozen or more times, they lost much of their charm. I could never decide if Old Frank's memory was slipping or if he merely enjoyed the act of telling stories.

And I suppose there was some valuable advice hidden somewhere in those stories. Especially if I ever decided to buy land during a Great Depression or wanted to sell fat hogs to the Department of Defense during the Second World War. But no matter what the content of his anecdote, he invariably ended it with "And I never made so darn much money in my entire life!"

I found that many of Old Frank's tales shared common themes. These included: Work hard; buy low, sell high; hang on to every nickel you can; and, above all, a little luck never hurt anybody. In fact, I believe that Old Frank was the luckiest man I ever met.

When he was about eighty-five, Old Frank decided to buy a newer, larger house. True to form, he eschewed the assistance of professional movers, choosing instead to press into service assorted acquaintances and neighborhood kids for the task.

When he was all moved into his new digs, Old Frank came to me and confessed that he had a problem. It seems that there was only one small item left to move that he and the neighborhood kids couldn't tackle: an ancient and massive Wells Fargo safe that inhabited his basement. I suggested that perhaps it was finally time to call professional movers, and he said that he already had.

"A hunnert dollars!" he exclaimed. "They want a hunnert dollars! And it's only two blocks! Why, they wouldn't even use a dime's worth of gas and they still have the nerve to charge an old man that much!"

I told him I didn't see any way around it. Old Frank smiled at me and said, "I know a guy who owns a pickup ..."

The next day, I found myself jockeying my pickup into position up against the landing that led to Old Frank's house. He had scrounged up some used planks to lay on the stairs and had borrowed a hand-operated cable winch that looked old enough to have hoisted Hannibal's elephants up the slopes of the Alps.

We put the planks on the stairs and threw a sturdy chain around the old safe. I attached the winch to my pickup, hooked the cable to the safe, and commenced winching. Old Frank insisted on staying below and behind the safe to supervise and to steer.

The rusty cable groaned and twanged under the tension as the safe inched its way up the planks. I expected something to give at any moment and anticipated the difficulty I would have in explaining why there was this very old and very flat man lying dead in his basement. But Old Frank's luck held, and the safe was soon resting in the pickup. My truck's rear end hung so low, I was afraid that its bumper would cut furrows in the driveway.

On my way out of the yard, I misjudged slightly and

nudged the neighbor lady's concrete birdbath with my bumper. The birdbath fell over and broke in two. Frank scowled and I empathized, saying, "Gee, that's too bad. I bet a new birdbath's gonna cost you about a hundred bucks."

No sooner had I said that than the neighbor lady stuck her head out the door and said, "It broke? Good! Will you haul that piece of junk away for me? I'd appreciate it."

But Old Frank's luck didn't end there. A few days later, he opened the safe to discover it contained a pile of old war bonds that he had forgotten. The interest they had accrued in the forty-plus years that had passed nearly paid for his new house.

Old Frank continued to come out to our farm almost daily. And every time he stopped for a visit, he would regale us with the same set of shopworn yarns, although his "raising fat hogs" story acquired a new twist. "And I put all the money into war bonds," he would conclude. "And I never made so darn much money in my entire life!"

My wife and I would simply smile and nod as we listened. After all, we had recently acquired—for free!—a slightly used, glommed-together-with-concrete-adhesive birdbath. And a colorful anecdote about how we came to own it. ●

The Lady Vet

~~~~~~~~~~~~~~~~~~~~~~~~~~~~~~

The doctor pressed her stethoscope against the cow's flank and thumped her finger against the bovine's belly as if she were checking on the ripeness of a gigantic, hairy watermelon. She had big brown eyes and weighed perhaps 110 pounds. The lady vet, that is; the cow, a Holstein, clocked in at about three-quarters of a ton.

"She has a displaced abomasum," murmured the lady vet.

Like any good dairy farmer, I knew that cows have a multichambered digestive system that is more complicated than the seating arrangement at a royal wedding. It had never occurred to me that a cow could somehow lose track of one of her stomachs.

"Not misplaced. Displaced," explained the lady vet. "Her abomasum has floated around to the wrong side. Nothing can get through, like when you twist a garden hose."

Eager to assuage the agony of my unfortunate ungulate, I asked about options.

"We could operate," said the lady vet, "or we could roll her."

It pained me to point out the obvious; namely, that the cow likely didn't own anything worth stealing. Lots of cowhide, but no wallet.

"What I mean is that we would lay the cow down and roll her over. Sometimes when we do that we can get the abomasum to shift back into place."

I replied that this plan sounded excellent except for the "lay the cow down" and "roll her over" parts. My experience was that it's extremely difficult to tip a cow, urban legend notwithstanding. Did I mention that the lady vet weighed perhaps 110 pounds? And I'm no Hercules.

"Not a big deal," she replied. "We'll just use the lariat."

The lady vet produced a lariat and looped it around the cow in several places, employing some mystical rope-tying technique that must have been handed down to her by a sage old cow shaman. When all was ready, she grabbed the rope and pulled. The cow responded by gently lying down.

Then came the exciting part. Cows do NOT enjoy being grabbed by their legs and rolled back and forth. Much cursing and sweating and grunting and dodging of flailing hooves ensued. Did I mention that the cow weighed three-quarters of a ton? And that I'm no Hercules?

At length the lady vet called a halt to our labors. She again thumped upon and listened to the cow's flank. Breaking into a grin, she said, "I think we did it!"

And so we had. We let the cow rejoin her herd mates, and she moseyed over to the bunk and commenced to munching hay.

I was deeply pleased. The cow had dodged the surgery bullet and I had learned that cow tipping is a real—and useful—activity.

I asked the good doctor what types of animals had come under her care.

"Snakes, rabbits, birds, you name it," she replied. "If it was on Noah's ark, I've probably treated it. Lizards. A camel. Water buffalo."

Impressed by her repertoire, I decided to put the lady vet to the test. I pointed to a nearby mother cat who was zealously grooming herself. The cat did this a lot; her life seemed to be an unending stream of bathing emergencies.

"Probably has fleas," said the lady vet. "Stop by my office and we'll give you a special shampoo for your cat."

Hearing the words "shampoo" and "cat" in the same sentence evoked emotions akin to hearing the words "nuclear" paired with "bomb." Even so, it seemed like an astute diagnosis.

Our farm mutt had been hanging around this whole time, supervising. I showed the lady vet a spot where the canine's coat had become thin and patchy.

"Could be ringworm," she said. "Stop by my office and we'll give you a special ointment."

I asked the lady vet if casual exposure to these ointments and shampoos would be hazardous for humans.

"No," she replied. "But it would likely clear up any fleas or ringworm you might have."

Striving to hide my sudden feelings of deep gladness, I asked the lady vet about some other symptoms. They included foul moods and intestinal bloating and cramping, which were inevitably followed by explosive eruptions that, reportedly, were loud enough to be heard in the next county.

"What species are we talking about?" she asked.

I admitted that the symptoms described were actually mine.

"It would be totally unethical for me to treat a human," she said. "But I could probably do a quick exam. Here, let's loop this lariat around you."

Don Quixote,
Tax Reformer

~~~~~~~~~~~~~~~~~~~~~~~~

He stopped by my farm again last August, the old knight. I had just finished another hot, sweaty day of field work and was headed toward the house when the thunder of hoofbeats announced his arrival.

He reined his steed to a halt a few yards from me and I took a good look at him; I hadn't seen him since the first Tuesday of the previous November. His armor was now dented and tarnished, and the grime coating it evidently the residue from some horrendous mud fight. Nevertheless, he was in high spirits.

"Salutations, good yeoman!" he boomed as he lifted the visor of his helmet. "I come today bearing glad tidings!"

I was instinctively leery because the last time he brought "glad tidings," he told me that I had just become the proud owner of some free grain. He neglected to disclose that it was stored somewhere on the Falkland Islands. I hoped he would go away soon, and yet I was curious. "What have you been up to now?" I inquired.

"Why, tax cuts, my good man, tax cuts! And a balanced budget to boot! I have it all here"—he dug into his saddlebag—"and I'm certain I have something which will be of interest to you!"

At length he extracted a fistful of papers. "Ah, here we are! Estate taxes! I have magnanimously increased the amount that is exempt from estate taxes! Doesn't that sound grand?"

"But wouldn't I have to die in order to take advantage of it?"

"Well, yes, yes, that's true. But I'm sure you'll agree that this is a relatively minor detail!"

"Even so, I think I'll pass."

The knight pressed on, unfazed. "Never fear!" he said as he continued to burrow into his saddlebag. "I have more! Much more! Here we go! Perhaps you would enjoy this superb investment tax credit?"

"Say what?"

"I will explain," he said as he looked around. "Ah! Say that you sell yon cow ..."

"Lulu Belle? She's not for sale. We use her as the family milk cow."

"Yes, yes. But say that you sell her. Some of the profits will be exempt via an investment tax credit!"

"But what would we do for milk?" I asked.

The old knight scowled for a moment, but quickly brightened, saying, "Why, of course! You could BUY milk with your tax savings!"

"Well, we've sorta become attached to old Lulu Belle."

"I see. Hmmm . . . this is a difficult case. Yes, most difficult . . . I have it! Are you the head of the household?"

I glanced around to make certain that my wife was out of earshot. "Yeah," I said at last.

"Good! And you have children?"

"Two boys."

"Excellent! You are automatically eligible for the increase of the per-child standard deduction! And if you have a child in college, you save even more! You simply fill out form 1827A, and if line 37 is less than line 26 and if your adjusted gross income is more than line 52 on Schedule WZ . . . By the way, do you recall what phase the moon was in at the time of your birth? Oh, never mind. Here."

He shoved both hands into his saddlebag and hauled out a bulky volume. "It's all explained in this," he said as he gave me the massive book. "These are all the changes to the tax code that were just enacted. It's all there in plain English."

I thumbed through the pages of gibberish. "Good

grief!" I exclaimed. "It'll take an army of accountants and lawyers to decipher all this!"

But the old knight was staring off into the distance. "Yes, yes," he said, distracted. "You can thank me later. I would love to dally but there is important business at hand. I must go now and slay yonder giant!"

He lowered his visor and hefted his lance. Spurring his mount, he cried, "Heigh-ho, Gridlock! Onward!" He galloped away, and ere he topped the hill, called out, "Remember me at the ballot box!"

"I will!" I hollered back. "I will!"

Just then, my wife walked out of the house. "Who was that?" she wondered.

"Oh, it was just that ditzy Quixote guy again."

"The congressman? What did he want this time?"

"Nothing. In fact, he was giving out tax breaks today. He claims they're all here in this big old book in some sort of secret code. I figure we can use it for kindling next winter."

My wife, ever the practical one, looked through it a moment and decided, "No, let's keep it. I bet it would be a big hit at the Annual Teeth-Gnashing and Hair-Pulling Festival next April. Now come on in, supper's on."

"In a minute, dear, in a minute. It's always fun to watch him prang off the side of that windmill."

Farm Supply
Stories

~~~~~~~~~~~~~~~~~~~~~~~~~~~

O ur town has one of those mega-huge farm supply
outlets. It's the kind of store where you will find
everything from panty hose to garden hoes to
hydraulic hose. It's a place where you can buy such things
as birdseed (if you want to grow birds) or duct tape (if you
want to tape ducks) or gravity wagons (if you need to haul
gravity).

A farm supply store is a mecca for farmers on rainy
days. If you were to step into such an establishment during
a monsoon, you might fall under the mistaken impression
that seed corn caps are required for admission.

You will see farmers—muddy boots and all—swapping stories, telling jokes, and generally having a farmer's
version of a jolly good time. Their wives will be participating in their own form of amusement, which usually
involves prowling the clothing department, hunting for
that ever-elusive clearance sale tag.

A farm supply store can also be a wonderful place

to take your wife on a date. This is especially true during Hotdog Days, when you can purchase a hotdog and a soda for pocket change. Your wife might not consider it a "real" date, but who can pass up a romantic rendezvous that costs only two bucks?

It makes you wonder how farmers managed to socialize before there were farm supply stores. When I was a youngster, we had this local event called a baling run.

The baling run that we belonged to consisted of half a dozen neighborhood farmers. One member was my dad's brother Coke, another was their uncle Stanley, and another was Martin, our Norwegian bachelor neighbor. These farmers all had a couple of things in common: They each had hay that needed baling, and they shared ownership in an ancient New Holland baler.

That old baler looked positively prehistoric. It was a massive conglomeration of gears and flywheels. Its main feature was a hay-packing apparatus that looked like a dragon's head. The packer mechanism would swing up and down with a rhythmic "chunka, chunka" sound, causing us kids to nickname the baler the chunka, chunka machine.

When the hay was deemed ready, a crew of neighborhood men, along with a fleet of assorted tractors and hay wagons, would assemble on the headland of the selected hay field. The men laughed and joked, filling the air with a carnival-like spirit.

As a little kid, I wasn't allowed to do much other than watch. Not that I didn't want to help. I would have liked nothing better than to throw those dusty, itchy bales of hay and to toil and sweat and spit like the men.

Sometimes the cottony clouds of morning would clump together and transform into the ominous thunderheads of afternoon. The flicker of lightning and the rumble of thunder would announce that Thor was swinging his mighty hammer against his anvil in the sky. The air would become so thick and oppressive that it felt more like a slurry than a gas.

A sweeping wall of rain would chase the baling crew from the field. The men would gather in the doorway of our barn, smoking and swapping jokes as fat raindrops exploded against the roof and the hissing downpour cooled and cleansed the sweltering summer air.

I clearly recall the smells of that old barn. There was the musk of sweaty men mingled with the fragrance of freshly baled alfalfa. The essence of cow manure mixed with the incense of cigarette smoke and the aroma of axle grease. It was wonderful.

It was during one of these enforced rest periods that

my uncle Coke first showed me that old "pull my finger" trick. Gales of laughter filled the barn at the sight of my surprised expression. I spent the rest of that day tugging on my finger, desperately hoping that I had inherited Coke's talent.

Many things have changed since then. Most farmers have their own balers or hire someone to bale their hay. Baling runs have largely become a thing of the past.

One summer afternoon when our two sons were young, the skies opened and it began to pour, so the family and I opted to drive to town and pick up some essentials. Our first stop was—of course!—the farm supply store.

As we entered the emporium, I said to our boys, "You know what this place needs? It needs some bales of hay so that all these farmers can sit a spell and feel more at home. Say, that reminds me of something. Come here and pull my finger." ⬤

Lutefisk Season

~~~~~~~~~~~~~~~~~~~~~~~~~~

Each year the holiday season brings with it all those holiday stresses and strains, like cleaning the house for the relatives, cooking banquet-sized meals for the relatives, and putting up with the relatives' bratty kids. Other than that, it is a relatively happy time of the year.

One of the biggest sources of stress around my household at this time involves the cultural chasm that exists between my German-blooded wife and Norwegian me: the cultural chasm called lutefisk. For the uninitiated, lutefisk is created as so: You start out with a big honking fillet of North Atlantic cod. As delectable as this might be, one must take a good thing and improve upon it, which is typical of the Norwegian mind-set. You age this cod fillet in the sun for a spell (unbelievers call it "letting it rot"), then steep the now stiff-as-a-board hunk of fish in a vat of lye water (infidels call it "soaking it in poison"). All that's left is to thoroughly rinse it in fresh water, boil it up, drown it in melted butter, and . . . mmmm! My mouth waters at the mere thought of this gastronomic delight!

During the centuries that preceded the advent of modern refrigeration, making lutefisk was an essential way to preserve calories for the winter. Folks who turn up their noses at lutefisk have probably never faced the choice of either slowly starving to death over a period of weeks or eating something that could quite possibly kill you today. Those of us who continue to enjoy lutefisk—even though lye-free fish is readily available at any supermarket—are honoring the memory of that long-ago Norseman who first said, "Yeah, sure, give me a chunk. I'll give that stuff a try."

And no Nordic winter holiday gathering would be complete without lefse, Scandinavia's answer to crepes and pita. If made correctly, lefse can be a tender and tasty addition to any meal. It can be used to mop up gravy or herd stray peas. Made incorrectly, lefse can be used for shoe repair. As with many culinary customs, there are numerous variations on lefse recipes, but they all have one thing in common: potatoes.

Lefse making begins with a kettle of boiled potatoes. The spuds are cooled and mashed and sent through a ricer to get rid of any lumps. Flour is then added. Most lefse recipes also call for sugar, cream, and some sort of shortening. The best lefse is made with lard.

The lefse dough is rolled out as thin as tissue paper, then given a quick turn on a hot griddle. A skilled, steady hand is needed during this process lest the lefse sheets become torn. The lefse is left to cool and is buttered and spun into rolls prior to serving. Lefse is an absolute necessity for cleaning one's plate of any extra melted butter or stray bits of lutefisk.

When my wife and I were newlyweds, we decided to host a holiday get-together for our families. On the appointed day, my wife was a blur of frenetic activity, scurrying about, cooking, cleaning, accumulating an aura of stress in the same manner that a powerful magnet attracts iron filings. I, on the other hand, was the epitome of coolheadedness as I sat calmly sipping a cold one while I simultaneously kept tabs on the score of an important televised sporting event. "Slow down," I advised from my perch on the recliner. "They aren't even gonna be here for a couple of hours! And then you know how it goes: Everyone's gotta stand around awhile and yak about how little Suzy has grown and on and on . . . I wouldn't even start boiling the taters until they got here."

My wife suddenly turned a whiter shade of pale. "Potatoes! I KNEW I forgot something. I gotta go to the store. Here, you stir the gravy while I'm gone."

She ran to the car and tore off in a spray of gravel, which reminded me of the fact that drag racing would be on later. I assumed command of the kitchen and made a quick appraisal of the situation. Everything seemed to be under control except for one glaring deficiency: There was no lutefisk cooking.

Luckily, I had taken the precaution of purchasing a massive slab of lutefisk a few days earlier. In short order, I had the lutefisk nestled happily in a pot of boiling water. When my wife jogged back into the house she stopped abruptly, sat in a chair, and worked up a crying jag. I gave her a hug. "I know," I said soothingly. "These holidays have a way of getting a person down."

"It's not that!" she sobbed. "Can't you smell it? Why did that stupid sewer have to back up today?" I told her that I couldn't smell anything amiss and tried to lift her spirits by showing her my culinary masterpiece.

"Gross," she said. "So that's what stinks. Don't tell me you're gonna eat this stuff. It smells more like it should be buried."

As our guests arrived, a definite pattern emerged. My relatives would enter our house, take a deep whiff, and say, "Mmm, smells like the holidays!" Her relatives tended to just wrinkle up their noses. Our relatives soon segregated into two groups: Mine gathered in the dining room and noshed the day away. Hers hung by the windows, making a big show of trying to suck in outside air.

Which all explains why I now have to cheat on my wife, so to speak. I wonder, can a guy thaw frozen lutefisk by holding it under his armpit? In any case, I don't imagine it'll hurt the flavor any.

Dear County Agent Guy

~~~~~~~~~~~~~~~~~~~~~~~~~~~~~~~~

I understand you're the one to talk to about herbicides. Well, I have a weed problem that no one seems to have the answer for. I am referring to cattails in my corn. I have looked at dozens of herbicide labels, and none are effective for cattails in corn.

While I'm on the subject of corn pests, maybe you can help me with another problem with my corn, namely ducks swimming around in it. I know those ducks are cute but their little webbed feet have claws on the toes and I'm afraid they might be shredding some corn leaves.

Another problem is town people. I don't mind them hanging out by my cornfield, but I wish they would leave them dang power boats and Jet Skis at home. Probably you know a county agent guy from way down South who could loan me a few of those gators that I hear they got. Maybe I could use them gators to take care of both the ducks and that other varmint problem at the same time.

It's been kinda wet this year. The other day, I was

driving up to the north place when I saw my neighbor standing out in his field in knee-deep mud. I stopped and asked if he needed some help. He said yeah, he was stuck. I told him I'd run home for a lariat. He said to forget that and bring my big chain. My chain? I asked. Yeah, he said, I'm standing on top of my pickup cab. I said I'd swing by and tell his wife that he'd probably be late. No need, he said, she's in the cab.

Isn't that dangerous? I asked. Why do you think I'm standing on the sunroof? he says. I ain't letting her out until I get a running head start. How's she taking it? I wondered. Well, he said, the continual swearing quit about ten minutes ago. Now he could just hear occasional reference to his low IQ and his recent ancestry. I said that ain't true, I know your mom and pop were married! And I know that neither of them was dogs.

Well, it happened to me, too. The other day, I got stuck. Not just a little, but buried to the muffler, hole in the world stuck. It took three tractors, four men, two kids, and a dog to get me out. (The dog was in charge of the life preserver. Or was it the brandy? No, that was last winter. Good boy, Elmo.)

So anyway, when they got me out, the neighbors and I stood and looked at the hole I left. It was almost as big as the tractors. One of them looked at it and said, "Well, that was fun."

"Yeah," I said. "It was almost as fun as the time I went to the dentist and he was trying to numb up the roof of my mouth and the next thing I know novocaine is rolling out of my nose. That was a blast."

Which got me to thinking about what I should do with that hole once it dried out. I thought about filling it in, but the kids said not until they practiced rappelling down the sides. So I thought maybe I'd advertise it as the northernmost branch of the Grand Canyon, but Fate took a hand.

That night, as I tried to sleep, I thought I heard a low moaning sound coming from that field. I got up to look, but darkness and fog had settled, and I couldn't see a thing. The next morning it was so foggy, you had to chew the air to get it down. I heard the low moaning like the night before so I lit out for the source of it. I got about fifty feet from where I got stuck, when the fog lifted enough for me to see a big something or the other had taken over my hole.

I stood there dumbfounded when a man hollered, "Ahoy there! We anywhere near St. Louis?" Sure as heck, there sat a big old barge full of grain in my field! Wet as it is, I hope for more rain so we can get that thing pulled out again. I mighta been stuck, but at least I wasn't lost, too.

That sorta reminds me of the time we made a trip to see the wife's uncle in California. One day her uncle Jim took me for a ride, and when we stopped we were by this big old slough. We stood there for a while looking at it until I couldn't stand it anymore.

"Jim," I says, "ya know, my friend Bud owns a 'dozer and a backhoe. I bet he could drain that dang slough!"

Jim says no, that might not be a good idea, that the folks in those parts had gotten sorta attached to the thing. He said they even named it, called it the Pacific or some such.

"Heck," I says, "it's no bigger than that lake up by Duluth! They gave that thing a snooty name, too—Superior, I think. Bud says in three, four days he coulda had that lake dry as a stove top. Probably could raise a heckuva crop down in there once a guy got it plowed. But folks up there wouldn't let him do it!"

Jim pointed out that the land here wouldn't be much good anyhow since the water in the slough was all salty. I thought about it a minute, then got an idea.

"Say, Jim, once you got it plowed, just plant it to peanuts! They'll come out of the ground salty, but that's how folks like them anyway!"

Just then, a couple of young gals came running along tossing a Frisbee back and forth. They musta got hot from running and giggling so much, and they started peeling off their sweat clothes. To my relief, they had swimsuits on underneath, but just barely. Those swimsuits were so skimpy that if you put them together, you could maybe wad a shotgun. While they splashed and giggled, I glanced around to see if there was a constable nearby. I was sure those gals was violating some ordinance, but since there was no cop around it looked like they would get away with it.

After a while the girls got tired, I guess, and spread towels out on the sand and laid down on them. I saw one gal slip her shoulder straps down and Jim said later that was so she wouldn't get tan lines there. I wish I had such things to worry about. On the way home, I says to Jim, "Ya know, maybe we should leave that slough alone. It's kind of a scenic place and I'd hate to spoil it." Jim agreed.

"How long do you reckon that water has stood there anyway, Jim?" He said he didn't know for sure, probably 250 million years or so.

"Well, there you go!" I said. " Everybody knows when the water stands more than a couple of years, it makes it harder than heck to plow. I think we'll leave it be."

Well, I'd better go. The weather service is predicting rain (again!) and our cattle yard is so wet, I'm thinking we'll soon have to buy scuba gear for the cows. I'll call you later about those cattails.

Part 2

How to Raise Farm-Fresh Kids in Twenty-Five Years or Less!

Labor and Delivery

An acquaintance of mine recently became a daddy for the first time. He talked about how the whole thing was some kind of moving, mystical experience. "Heck," I told him, "come back after you've been in on a couple hundred of those sacred events! They lose their charm after a dozen or so."

I was referring, of course, to livestock births.

When my wife became pregnant with our first child, I was forced into participating in something called "Lamaze" class. The alternative involved me sleeping on the couch. This is no fun, especially when you own a secondhand couch that has more lumps than a bag of potatoes.

I could see why they called it Lamaze right from the start. For one thing, I was L'amazed that the class set me back twenty bucks. The purpose of the class was also amazingly stupid. We were supposed to learn how to breathe. How many people do you know that don't already possess this skill? None, right?

During the class, we were to be instructed on how to

give birth. I had never once seen a cow in such a class. And so what if birthing can be difficult? Haven't those people ever heard of calf pullers? I pretty much knew all about that birthing stuff anyway. I'm pretty sure I could have taught the class myself. "Okay, class, listen up! When the time comes, the mama bellows and grunts until the baby comes out. If she has trouble, you can call the vet . . . I mean, the doc. Then she cleans it up while you have a beer. Any questions? That'll be twenty bucks, please."

Our class instructor turned out to be one of those high-strung "I know more than you so sit down and shut up" types. I think she sensed my resentment at being there (probably because I napped during the training films) and went out of her way to try to embarrass me.

During one session, she mentioned that the moms-to-be would probably hear horror stories from well-meaning friends and relatives about weeklong labors, babies coming out sideways, etc. Several ladies nodded, indicating that they had already heard such things.

"Shoot," I said, "I've been in on so many deliveries, I could tell you a few war stories myself."

The teacher raised an eyebrow and sneered at me like an executioner. "Perhaps you'd like to share one since you consider yourself such an expert."

"Well, okay," I began, "one time, I was doing my chores, and when I looked in the barn, I spied a new mom and her baby."

"Poor homeless people," murmured one of the gals.

I went on. "Well, the mother had cleaned the baby and was already nursing, so I—"

"I'm going to nurse, too," interjected another gal. "Natural is the only way."

I tried to continue. "So anyway, when I finished my chores I peeked in on her again and she had prolapsed."

The teacher became noticeably pale. "What's prolapsed?" asked one of the students.

"She had sort of coughed out her uterus." I saw several women wince in empathy. "So I called the doc right away and when he got there we lassoed her and tied her down—"

"You *what*?" said one of the ladies.

"Well, she wasn't going to let us help her. I guess she was scared for her baby."

The gals nodded with understanding. "The poor thing. I bet she couldn't speak a word of English, either."

"Yeah, that was pretty much a given." I noticed that the teacher was beginning to sway slightly. "Doc musta worked a half hour to get it all stuffed back in. He was real fussy, said he wanted to make sure there were no twists or wrinkles. Man, he had his arm in clear up to his shoulder, but that wasn't good enough, so he takes this quart pop bottle to use as an extension and he—"

My narrative was interrupted by a loud THUNK. The teacher had fainted dead away.

Class got out early that day. On the drive home, the wife chewed me out. "You should have told them you were talking about a cow. Sheesh! I've never been so embarrassed in my life."

"Yeah, but think of how the teacher felt. Did you see how her eyes rolled back? I was gonna shout 'TIMBER' but it was too late."

I was pretty sure we'd get an F in the class but it turned out we weren't even graded. I really didn't learn much new anyway. That is, not until the Big Event came.

Labor is a poor choice of a label for what goes on prior to delivery. It should really be called boredom. Hours and hours of hanging around a hospital room with a very pregnant (and often crabby) woman is not most guys' idea of a good time. But I was fully prepared for this contingency during the birth of our first child: I brought a deck of cards. Shuffling the deck, I said, "Hey, honey. How 'bout a game of strip poker?"

"No," she snapped. "That's how all this started,

remember? Besides, it wouldn't be fair, since all I'm wearing is this stupid hospital gown."

A nurse breezed into the room to check on things. Glancing at her name tag, I asked, "Hey, um, Nancy, how about a game of Texas Hold'em? By the way, are those shoes you wearing Reeboks? What size are they?"

My wife tried to hit me upside the head with a book of baby names. Luckily, she missed and was restrained from grabbing me due to all the tubes and monitors that were connected to her. I sought safety in a far corner.

After the nurse left, my wife asked innocently, "Would you check on this IV thing? I'm not sure if it's working." Eager to finally be of some use, I did as asked, and WHACK. She hit me upside the head with the IV pole. I couldn't believe that I fell for that old "check my IV for me" trick.

Morning dragged on into afternoon. Time seemed to stand still. I became so desperate for entertainment that I began sneaking peeks at a Clint Eastwood movie that was playing on a TV mounted high in one corner of the room. I had to be careful about watching TV, though; my wife wouldn't have liked my being distracted from the Big Event.

The nurse started to come in more and more often to check on progress. "How are we doing?" asked Nancy as she entered the room.

"Shh!" I said. "This is the exciting part! You see, Dirty Harry has the bad guy cornered, and I don't know if he's shot five times or six. I tried to count, but SOMEBODY"—I glared at my wife meaningfully—"decided to have a really noisy contraction just then."

It was at last determined that the time for delivery was very close. Nancy asked my wife, "Do you feel pushy?"

"When doesn't she?" I retorted. (I had long since safely moved all throwable items from my wife's reach.)

When the birth was finally at hand, I was told to dress in green scrubs and was admitted to the holiest of holies, the Delivery Room. The place was quite impressive. There were shiny electronic doodads and blinking thingamabobs everywhere. In the center of it all sat the delivery table, a tribute to modern stainless steel architecture. Imagine if a million years from now, an archaeologist discovers such a room: "This apparently was their torture chamber. Notice the metal instruments, both blunt and sharp, employed to inflict pain. See also the electronic equipment used to quantify the poor soul's discomfort. These were truly a barbaric people."

The atmosphere in the room grew tense as the minutes ticked by. I got the feeling that things weren't moving along as expected. I stood by, ready and able to be of service based on my experience with numerous farm animal deliveries. There seemed to be plenty of assistants, so I tried to lighten things up with anecdotes of some of the farm births I had witnessed.

"Yeah, I remember once when Dad hooked our John Deere B onto a calf that wouldn't come out, and he dragged that cow all over the place. And then there was that time the vet was doing a C-section on a cow and she busted out of her stanchion and we had to chase her down. What a mess THAT was. So anyway, the cow was running and Dad and I had to—"

"Surgical tape!" the doc barked tensely.

"Where do you want it?" asked the nurse.

"On the father's mouth!"

The doctor then picked up a pair of shiny little boat paddles. Before I could ask what they were for, he detailed how he was going to use the paddles to assist with the delivery. His description was quite graphic. I inexplicably broke out in a cold sweat, and my knees began to feel rubbery.

I glanced around to see if anyone else had been similarly affected. Nope. But everyone was giving me a strange look, as if I had just slapped on a witch doctor's mask and had begun to perform some sort of traditional baby-birthing dance.

But then I noticed something: There wasn't a cable winch anywhere in that entire room. What an outfit. Fortunately, I had come prepared.

"Hold on, Doc. Don't go and dislocate your shoulder. I got my calf puller out in the trunk of the car, just gimme a minute to go fetch it. I even hosed the puller off the last time I used it."

As the door to the delivery room closed, I heard a loud CLICK. They had locked me out.

Everything came out all right in the end. But I can only imagine how much quicker and easier things might have gone if they had just let me use my puller.

Electric Fencing 101

~~~~~~~~~~~~~~~~~~~~~~~~

C ity folks often ask me how it is to live in the country. I tell them it sure is great. Yeah, we don't have some of the more cultured things that you have in town. Stuff like the opera, cable TV, and that bizarre ritual called "rush hour" wherein drivers sit in their motionless cars and curse all the other drivers. Yep, we miss out on much of the glamour and the excitement of city life. But we have some things that you will never find in the city. A good for-instance would be electric fences.

The first cave-farmers probably had fences made of stone. We know this from cave wall paintings and *The Flintstones*. This meant that the gates were also probably made of stone, which would have posed a problem for the paleo-farmers. The gates proved too heavy for the cave-farmer's wife. This deprived him of one of a farmer's few pleasures—namely, blaming his wife for leaving the gate open.

For untold millennia, farmers experimented with different styles and types of livestock fencing. No real progress was made until the late 1800s, when Thomas Edison tamed electricity. One day, a farmer was prowling about the scrap heap behind Edison's lab when he came across an aborted attempt at a bug zapper. The farmer instinctively realized the huge potential of this technology and took the thing home.

The farmer hooked the zapper to a bare wire that was suspended by insulators between two posts. An old cow wandered up just then and touched the wire with her nose. When the farmer recovered from his flash blindness, he discovered two things: First, too much electricity was involved. And second, he had just invented a new method of barbecuing.

The farmer fiddled and tweaked with this new technology until hardly anything died when it touched the bare wire. Then, he tried it on his kids, inaugurating the time-honored tradition of farmers using their offspring as electric fence testers.

Many refinements have been made since then. Thanks to fallout from the Star Wars defense program, any farmer can purchase an electric fence charger with the innards of an electron beam gun. Now, the same power designed to obliterate intercontinental ballistic missiles can be used to keep your livestock in check.

Last spring, I bought such a fence charger. When it was all powered up, I called out our older son, a clever lad of nine years, and told him to bring a screwdriver. I showed the boy how, if you hold the screwdriver shaft against, say, a steel post and get the end of the tool to within a quarter inch of the "hot" wire, an electric arc will occur. I let him try it a few times, instructing him to take care that he touch only the insulated handle.

Later that day, the boy said he wanted to show me something. He took me out by the electric fence and called for the dog. The lad plucked a tick out of the dog's ear, placed the bug on the flat of a screwdriver blade, then proceeded to electrocute the hapless insect via the miniature lightning bolts that jumped from the fence. "Look at that!" he exclaimed. "See how the tick sorta vaporizes?"

The next morning, I was in the bathroom/library when I heard our younger boy run wailing into the house. It's a good thing his mother was there to interpret his blubbering. I assumed from the volume of his cries that his brother had amputated an arm during a scuffle. I instead heard my wife say, "Well, dang it! You shouldn't be frying ticks on the fence!"

I later buttonholed the older boy. "Did you show your little brother how to fry ticks?" I asked. The child admitted that he had. "You forgot to tell him to touch only the insulated handle, didn't you?" He said that it might have slipped his mind.

Kids! Who would have thought?

I have an acquaintance I'll call Harlan who shared with me an electric fence story. It seems that one autumn afternoon, Harlan espied an unfamiliar vehicle parked near the end of his pasture. In the distance, he could see a pair of trespassing hunters making their way out of his cornfield. Knowing that they had to cross the fence, Harlan switched on the electric fence charger, sending a kabillion volts through the top wire. Harlan was deeply gratified to hear a series of whoops and hollers roll across the pasture.

Minutes later, the vehicle pulled into his yard. To Harlan's surprise, the hunters were friends of his who were driving a borrowed car. Harlan said he was sorry, that an honest mistake had been made. "That's okay," said one of the men, "I was planning on a vasectomy anyway. Would you have any burn balm?"

Christmas Shopping
with a Caveman

~~~~~~~~~~~~~~~~~~~~

I was shopping with the wife last weekend when I saw a sure indicator of the upcoming holiday season: a mall zombie.

It's a certain sign that the Christmas season is just around the corner when you start bumping into these unfortunate men. Or what used to be men. Now, they are the living undead, brainlessly following their wives, pushing the shopping cart, a blank expression frozen on their faces.

What went wrong? How did these once-normal, red-blooded, beer-drinking, football-watching examples of manhood become such tragic wretches? The answer lies deep in human evolution.

15,000 BC: Og, your typical average cave dweller, is living the good life. He hunts and fishes all day while his mate keeps the cave clean. Her duties include gutting and cooking whatever Og happens to drag home. Og does no real work and pays no taxes; in short, he lives in paradise.

One day, the wind carries a bizarre new scent to Og's cave. Following his nose, Og trudges forth until he at last crests a hill. A bland concrete and steel structure squats in the valley, ugly as a toad. Emblazoned high on its fluorescent facade is the word "Mall."

Out of scientific curiosity, Og enters this mall place to investigate. He is shocked by what he sees.

Og perceives the masses worshipping the heathen god of materialism. He sees customers lined up at rows of tiny altars where priestesses accept the peoples' sacrifices. The miniature altars beep and ring with mechanical joy as they receive their offerings. He hears the faithful chant the mantra "blue light special, blue light special" over and over while seasonal Muzak drones in the background.

Og is disgusted as the shoppers pursue some elusive creature called a "bargain," which always seems to be up the next aisle or at the next store. He vows to quit this wicked place and never return, though he is sorely tempted by the everyday low, low prices in power tools.

A few days later, after a hard day of saber-toothed-tiger hunting, Og returns to his cave to find that a large something-or-other has taken over his living room. His mate enters, smiling.

"What that?" grunts Og.

"Isn't it great?" replies his mate. "We were going to get one anyway, and besides, it was on sale! I saved a lot!"

Og is pleased with the saving part. But later, Og is dismayed to discover that their secret trove of valuable animal skins has been considerably diminished. His instincts tell him his mate will never master the witchcraft of saving by spending, so he reaches a grim decision: Henceforth, he resolves, he must always accompany his mate to the mall, lest she "save" them to the poorcave.

So it is that the next weekend, Og finds himself at the pagan mall, tailing his mate while pushing a shopping cart.

Og notices a marked difference in the styles of male and female shopping. Lone male shoppers sprint down the aisles, their visits a mission, a purpose, a means to an end. He notes that they will zip to sporting goods and buy a dozen No. 6 spearheads before he and his mate even get to housewares. The shopping ritual seems to be its own reward for females. The women tramp from store to store, searching for the biggest bargain, probably so they can brag to their friends. Og can empathize with the bragging part.

The afternoon grinds on and on, and Og, overwhelmed by all the merchandise, lapses into a stupor. Unaware that he is drooling on himself, he staggers along behind the cart, his mind reeling with sensory overload. When the fog lifts, Og finds himself in an aisle that towers on each side with feminine hygiene products. The top layer of his cart is composed of these products.

Og is certain of three things: First, he is hopelessly lost. Second, all vestiges of his manhood have been stripped away. But third and worst of all, he knows he has missed the game of the week in its entirety.

Og's mate appears at the head of the aisle and scolds him for dawdling. He is belittled for not having a preference between mauve and teal. None of this bothers him anymore, for his transformation is complete: Og has become a mall zombie.

So this holiday season, as you cruise the parking lot in search of a spot near the entrance, watch out for these poor souls. Do not poke fun at their blank stares, their staggering gaits, the drool marks on their shirts. You may be next.

Never Sleep
with a Baby Chick

~~~~~~~~~~~~~~~~~~~~

I nearly made it big in the chicken business once. But then again, what does a four-year-old know?

My budding entrepreneurial career began when my parents went to town for "Hatchery Days." This was a gimmick the hatchery used to boost baby chick sales. Coffee and doughnuts were distributed freely, along with one day-old baby chick per customer.

My sister, then age six, and I were delighted when our parents presented us with this cute and fuzzy little bird. We called it Chickie, a name that seemed to suit it perfectly.

We were astounded at Chickie's intellect. At only one day of age, it already knew how to drink! It would also eat dry oatmeal right out of our hands. The hatchery man had foolishly given away one of the smartest chicks ever, we decided. We chose to keep Chickie's talents to ourselves and let his loss be our gain.

We began to make plans for Chickie's future. If it was a rooster, we would allow him the run of the house, patrolling for crumbs we kids spilled. This would save on chicken feed and housework at the same time. We would also train him to roost on our headboard to crow us awake each morning. More savings, having eliminated the need for an alarm clock. ("Waste not, want not" were words we lived by.)

If Chickie was a pullet, all the better. She would reward our love and care each day with an egg, which we would scramble and divide.

Then our plans grew. If Chickie became a mother hen, my sister said, she would someday have a brood of her own. Those chicks would grow up and raise their own families. As our imaginary flock increased, we envisioned truckloads of eggs and poultry leaving our farm every day. Then, said my sister, when the time was right, we would sell all and retire to the life of the independently wealthy. Our names would become associated with those of the Vanderbilts and the Rockefellers. About then, we calculated, I would have to start third grade.

Bedtime brought a problem. Chickie would be scared sleeping alone, we reasoned. We decided that she could just spend the night in the space between us. We tented the covers and Chickie seemed happy, walking around, scratching and peeping. We drifted off, dreaming of poultry profits.

When I awoke in the morning, Chickie was gone. I searched all around in the room, under the covers, beneath the bed. Gone. I woke my sister and shared the bad news with her. She sat up and rubbed her eyes and there was Chickie, lying on the mattress. Her flat, lifeless body looked so tiny and forlorn.

I scooped up Chickie's corpse and went straight to the authorities.

"Mom! Dad!" I sobbed. "Chickie's dead!"

The coroner examined the body. "Yep, she sure is," said Dad. "She looks kinda flat. Did you kids put her in the bed?"

I gave a tearful eyewitness account of the crime and pointed out exactly who the murderer was, but no arrests were ever made. Dad, always the practical one, suggested that perhaps we could salvage Chickie's tiny drumsticks and perhaps even make a wish with her miniature wishbone. It was years before I figured out that he was kidding.

Later that day, with a heavy heart, I went to the closet and located an empty shoe box. And then, tenderly, lovingly, I fed Chickie to a hungry mother cat. After all, waste not, want not. 🌸

Surviving
Parenthood

~~~~~~~~~~~~~~~~~~~~~~~

I don't mean to brag or anything, but my wife and I are fairly successful parents. And by "successful" I mean "our kids have never been featured on *America's Most Wanted.*" Well, not yet, anyhow.

Parenting is one of the strangest jobs in the world. Think about it: You don't have to do an interview, there is no test or license required, and you need no experience. In fact, the vast majority of us guys launched our parenting careers when we uttered a nearly universal two-word question: "You're WHAT?!"

Because parenting is basically a learn-on-the-job kind of job, we often tap into our own particular upbringings when it comes to bringing up the next generation.

When our younger boy was about four years old, he and I were sitting on the steps enjoying the afternoon sunshine. Pepper, our blue heeler (who was totally convinced that she was a people, too), sat between us. A childhood lesson suddenly washed over me, so I turned to the boy and said, "Blow in Pepper's ear." This he did, and the dog immediately responded by licking his cheek.

"She kissed me. Why did she do that?"

"It's automatic; she can't help it. All girls have to give kisses when you blow in their ears." I felt like a sure-enough parent by passing down this nugget of ancient wisdom.

The boy blew in the dog's ear again and got another slobbery kiss. He then trotted into the house, climbed onto his mother's lap, and blew into her ear. "What on earth are you doing?" my wife asked.

"You have to kiss me when I do that. Daddy said so." This struck her as sweet, so she smooched him on his cheek.

The boy was pleased. "Daddy was right," he said as he slid off his mother's lap. "It worked on Pepper dog and it worked on you."

My wife blanched. "You mean Pepper just kissed you? On that same cheek?"

A few minutes later, I found my wife washing her mouth with soap. "What happened?" I asked. "Did you catch yourself swearing?"

A bar of soap can really sting when it's thrown at you by an angry wife.

When you stop to think about it, parenting really

isn't all that difficult. For many of us, being a parent means simply imitating the things that were done by one's own parents. This is certainly true for me.

When I was a kid, Dad would, upon finishing a glass of milk, slam the empty glass onto the table and with much gusto exclaim, "Ah! Whiskey!"

I always thought this was a pretty cool expression, so when the time came, I taught it to my two little boys. It was cute when they first tried saying it and it came out as "Ah! Wicky!"

I don't know where Dad picked up that phrase, but would guess it had something to do with his stint in the navy during the Second World War. So it was that my boys were learning the mannerisms of some salty old sea dog Dad had once met.

As cute as it was to hear our little boys say that whenever they polished off a glass of milk, I was ordered to break them of that particular habit. My wife pointed out that it probably wouldn't go over very well with their preschool teachers.

I once met a farmer whose wife worked a full-time job in town. This farmer was somewhat of a stay-at-home dad and was therefore responsible for the potty training of their three kids.

Being a former farm kid himself, he naturally taught his kids—just as his dad had taught him—that it's okay to "go" outside so long as you ducked behind a tree or found a corner of a barn or some such.

He told me that this practice got him into hot water. "Our oldest was in kindergarten," he explained, "and was playing hard at recess when nature called. I had trained the kids to 'go' outside—and that's exactly what happened, right there in front of the teacher and the whole playground."

"You know what they say," I responded, trying to console him. "'Farm boys will be farm boys.'"

"True," he replied. "And so will farm girls."

The Ghosts of Horses Past

~~~~~~~~~~~~~~~~~~~~~

I t was one of my least favorite jobs, but I knew that I had to do it. The time had come to walk the perimeter of the pasture to inspect its fence and repair any damage the winter snows may have wrought.

I "volunteered" our younger son, Chris, to assist me. We set out upon our journey, each of us taking one side of the pasture. It was one of the first truly warm, sunny days of spring, and the pleasure of tramping across the grassy prairie nearly caused me to forget how much I disliked this chore.

The fence had taken the winter exceptionally well, and we finished our inspection tour ahead of schedule. Chris and I decided to reward ourselves by resting a spell on the rock-strewn escarpment that overlooks our pasture's slough. We rested on the knoll, soaking up the sun, enjoying the obscenely pleasant weather. We marveled at how it was that a long-ago glacier had hollowed out the slough while at the same time thrusting up this small, sharp rise—all for us to behold and enjoy on this outstanding spring morning.

I gestured toward the marsh. "You know," I said to Chris, "there's a lot of dead horses out there." He gave me that "Dad's lost it again" look, so I explained: Fifteen years ago, when his mother and I had just moved onto this farm, my father and I had embarked on this very same fence repair mission. This farm had been Dad's boyhood home. As he and I rested on this escarpment, he told me about the horses.

The thirties were a hard time, I explained. There was the drought, there were the dust storms and the clouds of grasshoppers. And then there was the mysterious disease that could strike the horses. I told Chris, as Dad had told me, how workhorses could be laid low by this ailment. Horses would often die of this malady, which was commonly called sleeping sickness. Those that survived were frequently rendered incapable of performing any useful amount of work.

This marshland was dry at that time, and since the digging here would be easy, horses that fell victim to this disease were often dragged out to this place to be buried. Graves the size of grand pianos were hand-excavated with picks and shovels.

Those horses were not merely tractors with hair.

They were more like huge pets, creatures who had been raised from birth, had been given names, and were known by their distinct personalities. The death of a good workhorse was more than an economic hardship; it was akin to losing a member of the family.

And back in those days, the losses were already high. For some farmers, losing a good horse to sleeping sickness may have been the final straw that pushed them over the brink and into the depths of despair.

We pondered this in silence for a long while, the quiet of the moment broken only by the chirruping of frogs down in the bog and the distant honking from a pair of Canada geese. Chris began to scout around for interesting rocks and found one that was flecked with fool's gold. We strolled down to the water's edge and chucked a few stones just to hear them splash and to help them complete their voyage to the bottomland, a geological journey that was halted ten thousand years ago.

My stomach told me that it was about lunchtime. On the walk back toward the house, we found a hole that some animal had dug and speculated as to whether or not a fox was building a den. We found a colony of red ants and used a stick to stir up their mound, then watched as the insects rushed around to repair the damage.

As we neared the house, I took a leisurely glance at my watch. My stomach clock had been fairly accurate, I noted with some satisfaction. But then I saw the date. What was it about that date?

Of course. Good God! How could have I forgotten? That day was the fourth anniversary of my father's death.

A wave of guilt swept over me. I could have—no, I *should* have—thought of something more fitting than checking the pasture fence to mark the occasion. But then I remembered that morning all those years ago when Dad and I had completed this same spring ritual, at a time when the strapping young man now walking at my side was still curled up in the muffled darkness of his mother's womb. I thought of the stony hill and of the story of the horses and how it had now been passed on to a new generation. And I realized that I could not have honored Dad in any better fashion.

Deep Diaper Doo-Doo

~~~~~~~~~~~~~~~~~~

I hate to brag, but I was the first to discover it. Back when our first son was just a baby, I stumbled across the ultimate antidote to masculinity.

So, what is it? What is this kryptonite-like substance that can make any man—even such paragons of machismo as Rambo or Norman Schwarzkopf—turn into blubbering sissies? Those of you who are married with children already know the answer. Soiled diapers.

What is it about baby poo-poo that can make a strong man weak and an average man lapse into a coma? I'm not sure, but I think it's because the stuff is nuclear.

Now, there are some guys who have the ability to change stinky diapers without so much as passing out. In fact, they don't even seem to *mind* this ungodly task! These men are not normal. They must have some sort of defect in their olfactory system; the same defect that allows women to soak themselves in perfume, yet somehow survive.

My wife was always disgusted with my aversion to changing soiled diapers. "I can't believe it!" she would say. "You're a dairy farmer! You work with tons of manure every day! What's the big deal about changing a messy little diaper?"

I tried to tell her that cow manure was different. For one thing, it wasn't radioactive. She whacked me upside the head with a box of Pampers. "Just go change the baby!" she ordered. "Sheesh! If it was up to you, that kid would be starting second grade still wearing the same diaper!" That's how *she* saw things. From my point of view, it was either him or me.

One day, my wife called me into the nursery, saying that she had something to show me. Her eyes held that twinkle that told me something special was waiting. Our son was no doubt gifted! Perhaps, at this young age, he had already mastered the Rubik's Cube! Or maybe he had perfected cold fusion, or was doing something *really* useful, such as engraving plates so we could print our own twenty-dollar bills!

I ran to the nursery, my mind racing with anticipation. There in the crib was our son—naked—and next to him was a diaper that had been filled well beyond capacity with a greenish goo.

The child had done a poo-poo. Not just a little one, though. No, this was a fill-up-the-diaper, *Guinness World Records* kind of poo-poo. The Mother of All Poo-Poos. My knees buckled, my brain reeled, and I broke out in a cold sweat. I suddenly didn't feel very well. This wasn't exactly the kind of talent I had hoped for!

"Look!" exclaimed my wife. "Have you ever seen anything like this?" My stomach rising to my throat, I croaked that I was trying not to think about it. Gathering the last of my strength, I staggered to the door. On my way out, I bumped the light switch and the room went dark.

And there in the blackness was a pale green glow!

The day before, the little guy had consumed a houseplant. Not the entire plant, mind you, but he fairly defoliated the thing and helped himself to some potting soil for dessert.

A frantic call to the pediatrician followed. After describing the houseplant our son had decimated, we were told not to worry, that the only result may be "a slight laxative effect." Riiight.

So when the news came out that the Department of Defense was in the market for a new antiterrorist weapon system, I wrote a letter to the Pentagon. I told them that the terrorists could be quickly brought to heel if we were to dump planeloads of soiled diapers on the bad guys' secret lairs. I soon received a terse reply. The generals had considered my plan, they said, and deemed it "a gross violation of the Geneva Conventions."

I showed the letter to my wife. "You men," she snorted. "You're such a bunch of wimps."

Uncle Wilmer

~~~~~~~~~~~~~~~~~~~~~~~~~~~~~

What do you say when a family farmer dies?

Is it enough to say that his rows were straight, that his furrows ran true? That he kept his fences taut and that you never saw any thistles growing upon his land? That he was a good neighbor in the biblical sense, one who was always quick to lend a hand in time of need? Small things, good things, all of those. But somehow, not nearly enough.

These questions came to the fore when my uncle Wilmer, Dad's brother, passed away. Like Dad, Wilmer died suddenly and unexpectedly while doing chores.

The next day, my wife and I drove the mile and a quarter east from our house over to Bev and Wilmer's place. As we approached their front door, we were met by Wilmer's rotund blue heeler. I stopped to pet the old dog and wondered who might scratch him behind the ears now that his master was gone.

We went inside. Awkward handshakes from Bev and Wilmer's sons, the now-grown men who, once upon a time, were the boys I spent summers with, playing 4-H softball. Gentle hugs from Bev and Wilmer's two daughters, their Nordic good looks reminding me of an old photograph of our youthful grandma Nelson on her wedding day.

Small things, good things.

Next came a big hug from Bev. Teary eyes and lumps in throats. But what to say? We all know that the day will come when we will have planted our last row of corn, when we will finish our final harvest. A day when we will feed our last cow or sow and climb out of that tractor or combine cab for the final time.

We who work with the rhythm of the seasons and the cycles of life know these things all too well. But none of that seemed to matter now.

I thought of telling Bev that Wilmer had lived his life as a good farmer should, communing daily with the Almighty beneath the great cathedral of the sky. That his family and farming were a part of his soul and that he would never be any farther away than the soil he so lovingly tilled.

It was as if Bev had read my thoughts. "Wilmer was doing what he loved to do," I heard her say.

A good thing, that. A very good thing indeed.

If there be such a thing as Farmers' Heaven, Wilmer is up there right now. I can imagine him sitting on the seat of his Super H on the headland of a field near the house. The vernal breeze wafts over him benevolently; the slumbering land stretches out before him like a canvas awaiting the master's brush.

Turning on the seat, Wilmer gazes back toward the farmstead. He can see his wife hanging clothes on the line while their young children play on the emerald lawn. The sun casts a golden hue and warms Wilmer's face.

Wilmer puts the plow into the ground, lets out the clutch, and throttles up. The plow bites into the rich black loam and slices a clean, sharp furrow. Wilmer smells the aroma of freshly turned earth and smiles broadly.

And he knows that this is a good thing. A very big and very good thing.

Dedicated to the memory of Wilmer A. Nelson,
December 24, 1927–January 24, 2001.

Husband Training
Made Easy

~~~~~~~~~~~~~~~~~~~~

O ur younger son was six years old and I was
schooling him in the fine art of advanced recoil
weaponry construction when I went to my wife's
dresser drawer to requisition some essential equipment.
To my surprise, I found a hidden book.

I wasn't surprised that my wife knew how to read. What
troubled me was the title of the book. It was called *How to
Make Your Man Behave in 21 Days or Less, Using the Secrets
of Professional Dog Trainers* and was written by Karen
Salmansohn. (This is an actual book that is still in print.)

I flipped through the tome. Several pages were dog-
eared, and I noticed there were numerous passages that
had been heavily underlined. I had just begun to thumb
through it when I heard approaching footsteps. I was trying to bury the evidence when my wife entered the room.

"And just what do you think you're doing?"

"I . . . um . . . you see, I was teaching the boy how to
build a slingshot and we were looking for something to use

for a sling, so I thought, 'Hey, why not make it a double-barrel?' and so we thought maybe we could borrow one of your Maidenforms..."

She swatted me across the nose with a copy of *Woman's World* magazine. "Get your grubby paws out of my underwear drawer! Bad boy! Bad, bad boy!"

I later got to thinking, an activity that is always fraught with hazards. How long had my wife had that book? And how had it affected our relationship? What about our boys? Were they also being "trained"?

Several incidents suddenly jumped into sharp focus. I remembered how, while potty training our boys, my wife had lined the entire bathroom with newspapers. And that time we visited the big city, it was she who insisted that both our boys be put on leashes.

I considered the suppertime ritual at our house. Up until now, I had thought it was cute the way she had conditioned the boys to come running at the sound of the can opener.

As I rolled these things over in my mind, I became incensed. What did she think she was doing? Running a kennel? The thought raised my hackles. I went hunting for my wife, intent on chewing her out. Bounding into the kitchen, I growled, "I have a bone to pick with you!"

My wife turned away and waved her hands at me. "Egad!" she said. "Your breath is horrible! Have you been drinking out of the toilet again? Here, chew on this. Now, what were you saying?"

"It's that book!" I said, munching on the Milk-Bone. "You've been treating the boys and me like . . . like . . ." My train of thought evaporated as she dangled a luscious strip of bacon in front of my nose.

"Do you want this, boy? Do you? Then go fetch my slippers! Go on! Fetch, boy!"

I reeled my tongue back in and wiped the saliva off my chin. "Wait a minute!" I barked. "I'm not falling for that old trick again! As I was saying . . ."

She came over and began to scratch me behind the ears. I felt becalmed as she spoke to me in a low, soothing voice. "There, there," she cooed, "let's not ruffle your fur. By the way, do you remember what day this is?"

Uh-oh! I knew what that meant! I tried to slink from the room. She caught me by the collar and began to drag me in the direction of the bathroom. I struggled and yelped and whined.

"Oh, be quiet!" she commanded. "I don't like this any more than you do! Come on! You know darn well it's time for your bath!"

Later, I sat on the front steps and reflected upon the day's events. Our youngest boy came out and plopped down beside me.

"Mom's mad at you, isn't she?"

"Yep. I'm in the doghouse again."

"She really yelled when you shook yourself and got that flea shampoo in her eyes! I bet she won't let us finish that slingshot now."

"Probably not."

"That's too bad. It would have been the ultimate anti-cat weapon."

We shared a moment of sullen silence. Then the boy said something that always makes my ears perk up.

"Hey, Dad, look! A squirrel!"

Monkey
Business

~~~~~~~~~~~~~~~~~~~~~~~~~~

When our nephew Adam and his wife, Janine, had their baby boy, it was a momentous event for our family. Ayden is the first child to be born to the next generation and my mother's first great-grandchild.

We wanted to mark the event with a gift, but had few ideas regarding what the new parents might want or need. We opted to send them a toy stuffed monkey, along with the following toy monkey story we wrote to go with it:

The summer when I was seven years old I met a red-haired eight-year-old boy named Steve.

Steve was spending the summer at our neighbor's farm, and I was brought in as a potential playmate for him. Within moments of our meeting, Steve suggested that we go out to the grove to climb trees. This caused me to take an instant liking to him. Our friendship—which has now spanned more than four decades—was cemented by a story he shared while we were up in the first tree we climbed together.

What he told me was a joke, perhaps the first one I had ever heard. It went something like this:

There was a hog farmer who wanted to win the prize for the biggest pig at the county fair. He decided to cheat a bit by forcing a cork into the hind end of his largest hog. The pig quickly grew heavier and heavier until it became unbelievably huge.

The farmer took the hog to the county fair, where it was judged to be the biggest pig. The judges were just about to give the farmer the trophy for his hog when an escaped circus monkey scampered into the arena.

The monkey saw the cork and decided to give it a tug. After the ensuing pandemonium died down, a newspaper reporter interviewed eyewitnesses.

"What did you see?" he asked one of the judges.

"Poop flying all over!" he replied. (At this point Steve extravagantly pantomimed mopping his face, which caused me to chuckle.)

"What did *you* see?" he asked the second judge.

"Poop flying all over!" (Again with the wiping of the face. Many more giggles.)

"And what did you see?" the reporter asked a bystander.

"Monkey trying to get the cork back in!"

I nearly fell out of the tree due to massive paroxysms of laughter. The image of the hapless little monkey struggling to replace that cork in the midst of a fire hose–like blast of pig poop was burned onto my mental retinas. It remains there to this day.

For a seven-year-old boy, it just doesn't get any funnier than that.

Flash forward more than twenty years; I am the father of two little boys of my own. It has become a ritual for me to read them bedtime stories, which I am glad to do. It gives me an excuse to expose them to such literary triumphs as *Calvin and Hobbes*, *The Far Side*, and the works of Patrick McManus, the author of such cerebral gems as *Never Sniff a Gift Fish* and *Real Ponies Don't Go Oink!*

One night, we found ourselves short of reading material and the boys were clamoring for something more. On sheer impulse, I grabbed a toy stuffed monkey named Monkey and retold the pig joke with Monkey pantomiming the part of the monkey.

The boys giggled themselves silly. "Do another!" they exclaimed at the conclusion of the joke. Still running on impulse, I began, "One day, Monkey was walking along when he found a stick of dynamite. 'Wow!' said Monkey,

'This is my lucky day! How often does a guy find a big, fat cigar like this? Now, if I could just find a lighter . . .'"

Monkey went on to have similar misadventures nearly every night. Most of them involved substances like superglue, nitroglycerin, and oil of ipecac, and very often a combination of ingredients.

And now, Adam, you are a brand-new daddy. This job comes with many responsibilities, not the least of which is telling bedtime stories. It may be a while before your little boy begins asking for stories, but that time will arrive before you know it.

Enclosed is a toy stuffed monkey to help you get started. It will be up to you and your little guy to decide what sort of adventures the monkey might have and what kind of messes he might get into. No matter what they may be, they will most certainly be special.

The enclosed monkey is not the original Monkey I used to entertain our boys. That Monkey is enjoying a quiet retirement high up on a closet shelf in Christopher's room.

After all, he deserves it. Especially after that time when he mistook a flamethrower for a Waterpik. ●

I'm Gonna Marry Mrs. Mortimer!

~~~~~~~~~~~~~~~~~~~~

Everyone remembers the first time they fell totally and hopelessly in love. I remember my first as a period of quiet times spent together, holding hands in public, thrilling to that "special" look across a crowded room.

My third grade teacher, Mrs. Mortimer, was an angel.

Ours was no ordinary love, but then Mrs. Mortimer was no ordinary woman. Her love was so powerful, she was willing to risk her very life for me.

I don't recall the exact moment when I first realized that I had special feelings for Mrs. Mortimer. I think it was when she selected me—much to my classmates' great envy—to empty the wastepaper basket.

I soon began to actually look forward to attending school. And this from a kid who had previously spent school mornings under the couch, dodging the broom handle that Mom was using to flush me out of my hiding place.

Once I comprehended the seriousness of my condition, I knew that there was only one thing a guy could do: Mrs. Mortimer and I must be married. I shared my plans for the future with my pal Ernie one day while we ate lunch.

Mrs. Mortimer "happened" to walk past our table and gave me one of her special smiles, a tiny grin that caused my heart to swell with unbounded joy. When she was out of earshot, I said to Ernie, "I'm gonna marry Mrs. Mortimer!"

"You dope," Ernie said. "You can't marry her."

"Why not?" I said. "She's a woman, and I'm . . . Well, I'm gonna be a man someday."

"That's not it, dummy. It's 'cause of her first name: Missus. That means she's already married."

I was devastated. I was so depressed that when Ernie dumped his peas into his milk and said he was drinking newts' eyeballs, I could barely manage a weak giggle.

As it is with that crazy roller coaster called love, one day I was given a ray of hope. Someone had seen Mrs. Mortimer's husband, and reports were that he was old. His age was estimated at thirty-five, perhaps even forty. This was encouraging news indeed. Why, the man had one foot in the grave. It would only be a short while before my favorite teacher would become a widow. And who would be there, ready to give solace, eager to empty the wastepaper basket? You guessed it.

I had no idea whether Mrs. Mortimer had the same feelings for me that I had for her. That is, not until that fateful day when she laid down her life for me.

It was a warm spring afternoon and I had been playing hard at recess. I found it necessary to remove my coat and decided to hang it in our classroom closet. This wasn't because I was any sort of neat freak, but mainly so that I could catch a glimpse of my ladylove as she corrected papers.

I lingered at the doorway, drinking in the sight of her as she marked math tests with her red pen. But I tarried a bit too long and the automatic closer shut the door on my fingers.

Mrs. Mortimer rushed to my side at the sound of my yelps. She inspected my hand and, without hesitation, spirited me off to the boys' bathroom. "Oh, no!" I thought as she led me over to the sink to run cold water over my injury.

We boys had been told (by the girls) that if a boy were to enter the girls' john, his head would instantly spin around and explode. That's if he was lucky. If he was unlucky, the rest of his brief life would be turned into a living hell by hordes of girl cooties.

I assumed that it worked both ways, so I looked up at Mrs. Mortimer for what I thought was the last time. The tears I shed weren't just from pain, but also for the heroic woman who loved me so dearly that she was willing to make the ultimate sacrifice.

But then . . . a miracle. Her head didn't explode. No doubt her love for me was so strong that she was able to overcome the effects of boy cooties and save my life! Such a profound and selfless love comes along perhaps once every million years.

After third grade, Mrs. Mortimer and I drifted apart and went our separate ways. I have no idea what became of her.

If you're out there, Mrs. Mortimer, please know this: There is a certain former third grader who will empty your wastepaper basket anytime.

Staying Married to a Dairy Farmer

~~~~~~~~~~~~~~~~~~~~~~~

I recently received a letter from a man named Wilfred who mentioned that he is eighty-eight years old. The envelope bore only my name and hometown, but arrived intact and on time. Try *that* with email. Wilfred said that he and his wife, who is a mere eighty-seven, have been married for sixty-five years. There aren't many who can truthfully say that their marriage license would be eligible for Social Security.

I wrote back to Wilfred and congratulated him and his bride for their long marriage. I also noted that there are probably some who are still saying, "Yeah, well. I bet it won't last."

Marriage is a dicey proposition at best. Half of all unions end in divorce court, while the other half end at the funeral parlor. Neither prospect seems very pleasant. And it doesn't help that supermarket tabloids spew a torrent of gunk about movie star hookups and breakups. What? You mean to say that marriages made in Hollywood don't last? Who would have thought?

My wife and I have been married for over thirty years, which means we're still on training wheels compared to Wilfred and his wife. But even at this early stage, we've managed to accumulate some insights regarding what it takes to make it last. Here are a few tidbits:

Be prepared to compromise.

One evening, when my wife and I were newlyweds, she asked, "Do you suppose we could go out on a date tonight? You know, dinner and a movie?"

I pointed out that the whole point of being married was that a person would quit dating. Besides, as struggling young dairy farmers, we could scarcely afford such an extravagance as movie tickets. We had a perfectly good TV that received four perfectly good channels.

My wife was insistent, so I compromised by taking her out for dinner and a movie. When we returned home to our little farm, we were greeted by a swarm of flashing red lights.

It seems that our cows had noticed our departure and used the opportunity to escape their pen and go galumphing about. Our farm was located next to a busy highway, so the authorities were summoned soon after our Holsteins made their first attempt at jaywalking.

We both knew that I was to blame for this debacle, because I was the one who hadn't properly secured the gate. My wife compromised by not mentioning this ever again and I compromised by continuing to take her out on dates.

● **Be sensitive to each other's feelings.**

A good example of this was when my wife was pregnant with our first son. She was quite apprehensive about the whole situation, while I was the epitome of cool, calm collectedness.

This was because I had witnessed dozens of farm animal births and had reached the conclusion that A) it isn't that big a deal, and B) it don't hurt all that much, and C) should difficulties arise, there are numerous surgical and mechanical methods that can be used to address them.

My wife didn't care for my hilarious anecdotes regarding difficult farm animal births and told me so in no uncertain terms.

The birth of our first son was indeed difficult. Being sensitive to my wife's feelings, I didn't mention during the height of the crisis that someone had thought to pack the calf puller and that it was at the ready in the trunk of the car.

● Be nice to each other.

You wouldn't think that this even needs to be said, but it appears this isn't the case. Many couples seem blind to the fact that kindness is essential for a long and happy relationship. This should be common sense, much like "never pick up a white cat when you're wearing a black sweater" or "don't throw a lit firecracker into a puddle of gasoline." Being kind to each other is just that easy.

Some years ago, I decided to try the sport of skydiving. Did my wife insist that it was lick-a-frozen-flagpole foolish to leap out of a perfectly good airplane? Nope. She was kind and indulgent, never once invoking the old adage that there are only two things that fall from the sky: bird poop and idiots.

Things went haywire during what would be my final skydive. I lost control during free fall and my universe became a spinning blur of earth and sky. The parachute opening was a hard one, leaving bruises that made it appear that I had been on the losing end of a battle with a sledgehammer. The parachute lines became twisted, causing an "I'm gonna need clean underwear!" feeling to well up inside me as I dangled beneath the canopy at two thousand feet.

After I somehow managed to pull off a safe landing, my jumpmaster took me aside and advised, "There are old skydivers and there are bold skydivers, but there are no old, bold skydivers. You're not doing so good with the bold part."

I decided then and there that I'd had enough of skydiving, and my wife has been kind enough to never mention that it was a birdbrained idea to begin with.

These are just a few of the pointers that we have picked up over the years. And while we may never match the track record of Wilfred and his wife, we take comfort in the knowledge that we're definitely going to outlast all the Kardashians. ●

Hawaii Boy

W e all evolve as we travel through life. For example, our younger son was called (by me) Annoying Boy when he was a teenager. This was changed to College Boy when he started at the university, but that was soon modified, for obvious reasons, to Laundry Boy.

And then, in the second year of college, he acquired an entirely new and unexpected handle: Hawaii Boy.

The driving force behind this strange evolution is the fact that South Dakota's university system has reciprocity with the University of Hawaii at Manoa. When I learned of this, my initial reaction was "You must be kidding!" When I found out that they weren't, I asked a person I know at South Dakota State University if I could get in on some of that reciprocity stuff. I was told that I could but that I would first have to enroll and become a full-time student. Nothing spoils a bunch of prospective fun quicker than being told that learning is part of the bargain.

I asked if any Hawaiians take advantage of this deal. I was shocked to learn that some actually do! I think that Hawaiian students should pay extra, as there is nothing like the pure, unadulterated misery of a Dakota winter to teach one how to fully appreciate life in paradise.

My wife grew ever sadder the closer Hawaii Boy's departure came. "Don't worry," I assured her, "he'll be back someday. I guarantee it."

"How can you be so sure?" she asked, wiping away a tear.

"Because sooner or later, he's gonna run out of clean laundry."

~~~~~~~~~~~~~~~~~

My wife and I learned, via Hawaii Boy's emails and phone calls, that his education was being extended far beyond that which was taught in the classroom.

For instance, he discovered what it's like to be in the minority. In Hawaii, tall blond males are somewhat of a rarity. On the other hand, he was offered a job at Abercrombie & Fitch minutes after walking into the joint.

Hawaii Boy also found that wildlife abounded in paradise—some of it unexpected. He once complained to us that a pair of mongooses (mongeese?) had spent the night fighting beneath his window. Mongooses were brought to

Oahu over a hundred years ago to control rats in the sugar-cane fields and are now as much of a pest as the rats.

~~~~~~~~~~~~~~~~~~~~~

A fter a few weeks, my wife and I decided that it would be cruel to make Hawaii Boy spend Christmas without us, so we purchased tickets that would wing us to Honolulu. The cost? Let's just say that my credit card emitted a warm nuclear glow for several days after the purchase.

But this visit was to be a twofer for us. Our nephew Adam, who has been like a son to us and a brother to Hawaii Boy, is an army nurse. Adam had recently been assigned to a duty station on Oahu. I imagine the news was delivered something like, "Listen up, soldier! You are being assigned to a new duty post. The work is going to be difficult and the hours will be long. Oh, and by the way, your new station will be located in the midst of paradise."

~~~~~~~~~~~~~~~~~~~~~

I t always drove my wife and our kids crazy whenever we traveled somewhere. I would gaze up at the craggy visage of Mount Rushmore or gawk at the lunar landscape of the Badlands and ask, "How can they grow corn here? Or even wheat?"

Growing nearly any crop doesn't appear to be a problem in Hawaii. With 365 frost-free days (although we espied snow on the summit of the slumbering Mauna Kea volcano), the only limiting factor for plant growth would be rainfall. Annual precipitation in Hawaii varies widely from place to place. The windward side of an island can average over three hundred inches of rain per year, while the leeward shore might receive less than ten. This difference can be so stark, it's been said that you can lean a double-barrel shotgun up against a post and one barrel will overflow with rainwater while the other will remain dry.

The island of Oahu rises up from the depths of the Pacific like a vast sea monster with numerous undulating humps that pierce the low-lying clouds. These jagged, emerald-hued peaks tickle the clouds' fat bellies, causing them to laugh until they cry.

Every morning in Hawaii smells and feels like the first real day of spring. Birds—mostly mina birds and some sort of small gray dove—sing from the treetops. You wake up and walk out and it's seventy-three degrees at sunrise. The high today will be eighty. The low tonight will be seventy-three. Throw in a fifty percent chance of rain and repeat into infinity.

The aroma of flowers and freshly cut grass is everywhere. Lawn mowing is a never-ending task on Oahu; on the other hand, they never have to shovel snow or scrape windshields.

We stayed at Adam's house, which was neither heated nor air-conditioned; I doubt that it was even insulated. Open windows catch the perpetual trade winds for nighttime cooling. A person could probably sleep outside on the ground if he didn't mind waking up to a wet bed due to nocturnal rain showers.

Honolulu is a bustling megalopolis, complete with gleaming skyscrapers, stiletto-heeled streetwalkers, and six-lane traffic jams. It takes two people to drive in Honolulu: one to steer and one to man a giant shoehorn whenever you want to commit the near-suicidal act of merging. Surprisingly, no one ever uses their horn.

It can rain at any random time of the day in Honolulu, even when the sun is shining at full wattage. As a result, rainbows are as common in Honolulu as roadkill is back home. Rainbows are such a familiar sight, Honoluluians pay them scant attention, no doubt thinking, "Ho-hum. Just another rainbow."

In the Midwest, rain and rainbows are minor miracles, something to celebrate. These little miracles elicit only yawns in paradise.

Despite being the most urban of the Hawaiian islands, Oahu is also home to a good number of wild chickens and feral pigs. Knowing that such familiar livestock was present was a source of comfort for a former farm kid. Some of the hogs aren't all that wild, though, and have been seen strolling along Waikiki Beach wearing much-too-small Speedos.

Adam's house was situated just two blocks from the beach. The incessant trade winds cause sand to drift from the beach, blocking beachside footpaths. Were it snow, a person could think, "It'll melt in a few days, so I don't think I'll shovel it." I'm betting this is not an option with sand.

～～～～～～～～～～～～～

During my years as a dairy farmer, I acquired a knack for tackling difficult projects for which I had no experience or training. This ability came in handy during our visit to Hawaii.

Adam and Hawaii Boy insisted that we eat at a restaurant that specializes in a bizarre substance called "sushi." I had no idea what to order, so I asked Adam to pick something from the menu for me. The impish gleam in his eye caused me to wonder what I might have gotten myself into. At length, a plate was set before me. Upon it sat a wad of rice wrapped around a piece of ichthyoid flesh that could best be described as "undercooked." The fish was, in fact, completely and totally raw.

"Go ahead, try it," urged Hawaii Boy in a tone of voice normally associated with tongues frozen to flagpoles. I had little choice in the matter. After all, I am an avowed eater of lutefisk, which is essentially poisoned cod that has been left out to dry. Closing my eyes, I popped the sushi into my mouth and chewed it down.

"Not bad," I said. "Tastes like fishy rice."

"Now try some wasabi," challenged Adam. A small dollop of an evil-looking green paste was smeared onto the edge of my plate. I put a dot the size of a pencil eraser on my tongue. A mistake. Wasabi, it seems, translates to "nuclear horseradish." My innards immediately began to protest as they launched into a full-blown meltdown.

~~~~~~~~~~~~~~~~

H alfway through our weeklong tropical odyssey, we decided to take a two-day jaunt to the island of Hawaii. One afternoon, while motoring around the Big Island, we were all struck with severe hunger pangs. We stopped at a small convenience store to pick up some munchies, which turned out to quite different from those found here in the Midwest. I went straight to the beef jerky section, only to discover almost no beef jerky. My wife grabbed a bag of smoked ahi—a form of tuna, which tasted like fish—while I considered my options. I eschewed the smoked cuttlefish legs, instead choosing a bag of smoked . . . something. Only after consuming several bites of the chewy stuff did I look closely at the label.

The first ingredient: squid. It wasn't bad, though. It sort of tasted like fish.

The North Shore of Oahu is dotted with roadside food stands, many of which sell locally grown fruit. We stopped at such a stand to examine the offerings. I would probably just mangle a mango and figured I had best stay away from the passion fruit, since I'm a married man. What interested me most was the pineapple.

I asked for a pineapple sample and was handed a golden wedge. My unsuspecting taste buds nearly had a coronary. The ultrafresh pineapple was sweet and tender and only vaguely similar to the canned stuff I had consumed all my life. It was so good, I nearly cried as the sugary juice dribbled down my chin.

And that's too bad, because now I'll probably never

again be able to stand canned pineapple. Consuming the fresh Hawaiian stuff has turned me into a pineapple snob.

~~~~~~~~~~~~~~~~~~~~~~~~

On our final night on Oahu, we all went to an establishment called Pinky's Pupu Bar and Grill. Hawaii Boy and Adam again issued a challenge to me, this time with something called an "oyster shot." A tall shot glass was placed before me. In it sat a raw oyster, topped with a dollop of cocktail sauce.

I tossed the contents of the shot glass into my mouth and began to chew. "Gross!" exclaimed Hawaii Boy. "You're not supposed to chew! You're just supposed to swallow!"

"How was it?" asked my wife after I'd finished.

"Not bad," I replied. "It sort of tasted like lutefisk."

Part 3

Never Kick
a Fresh
Cow Pie:

Lessons Learned
from a Lifetime of
Dairy Farming

A Dairy Farmer's Vacation

~~~~~~~~~~~~~~~~~~~~~~~~~~~~~~~~

he aroma of warming dirt is like espresso to a
farmer: His pulse races, he becomes wild-eyed,
and his hair stands on end. This is why farmers
wear seed corn caps.

Some of the land that Dad and I farmed lay next to a
highway. Sunny spring days would find us frantically rushing to get our crops planted on time. Off in the distance, I
would see campers and boats gliding by on the highway and
wondered how those folks could do that.

What was wrong with them? Hadn't they ever heard
the expression "Make hay while the sun shines and early to
bed early to rise and Rome wasn't built in a day and if you
chop your own wood it warms you twice and idle hands are
the devil's workshop"?

On one such sunny spring day, Dad and I were tilling a
field to prepare it for corn planting. As I turned my tractor
at the end of the field, I saw that Dad's tractor was sitting
on the opposite headland. He had stopped work and had

climbed down from his tractor cab and was gabbing idly with the neighbor, as if that field of corn were going to plant itself.

I drove to the headland to see what was so important that Dad and the neighbor had to discuss it for half an hour. Just local gossip.

I later mentioned this to Dad, and he said, "You should rest the horses at the end of the field." I said his "horse" was made of steel and didn't need any rest.

"I wasn't talking about the tractor," he said.

He had a point. A person shouldn't be so consumed by making a living that he doesn't have time to live. All work and no play sounds suspiciously similar to a chain gang.

When I was a kid, Dad would make time at least once a summer to take us kids on fishing expeditions to a local lake. We had to be back home in time for evening milking, which means we originated the daycation.

The morning of the fishing trip would find us rushing about like kittens in a roomful of yarn balls. There were preparations to make—the main one being casting practice.

Our family owned exactly one fishing rod. It was equipped with an open reel, the kind that would snarl into a humongous knot on almost every cast. It took a while for each of us to get just one practice cast.

We believed that the largest fish lived in the deepest water, which meant that casting farther equaled bigger fish. This is why we used a large square nut, which weighed half a pound, as a sinker.

Shortly before we left for the lake, a couple of us kids dug for worms. As we dropped the slimy, slithery, subterranean invertebrates into a coffee can, I wondered about the wisdom of eating something that would eat such a thing.

We piled into our 1959 Ford station wagon, drove to the lake, selected a likely spot on the shoreline, and commenced fishing operations. The ickiest part—baiting a razor hook the size of a crowbar—came first. None of the worms seemed pleased that they had been chosen to play a central role in our excellent adventure.

The first cast was made, and the reel instantly snarled. When we finally got it untangled, we saw that the bobber was bobbing. The fishing line was spooled in and there was a bullhead on the hook! Woo-hoo! We fished that spot until we caught a dozen bullheads. Some were about four inches long, but we caught some small ones, too.

I recalled this pleasant experience many years later when our sons were in grade school. Despite the fact that it was a wondrous spring day and we had *way* too many things that needed to be done around the farm, I decided to take the family fishing.

My wife and I took the boys to a local discount store and outfitted them with new rods and snarl-proof reels. Live and artificial bait was purchased. Standing on the shore of a local lake, we made numerous soaring and tangle-free casts.

But nothing. After drowning dozens of minnows, we switched to night crawlers and finally began to catch fish. What began as angling agony ended with fishing euphoria. All the fish were bullheads, of course. But we also landed some very excellent and sun-soaked memories.

Over the years, no matter how much stuff we had to do on our dairy farm, we tried to keep alive Dad's practice of taking quick little vacation-like daytime jaunts. My wife and I have become practiced "day-trippers."

As with most activities, being a day-tripper is a skill that requires constant honing. For example, one summer weekend some years ago, our youngest son, Chris, had a friend over for the weekend. Having a couple of fourteen-year-old boys around the house makes you feel like the guy who has been put in charge of the young carnivores at the zoo. We are talking about critters who are in perpetual motion, holding mock battles, testing each other's strength, and pacing the floor when they become bored.

"That's enough!" I announced. "You boys are wearing out our nerves, not to mention the carpet! Everybody outside and into the car!"

We herded the boys into our family sedan and set out with no particular destination in mind, just so long as it was fun, cheap, and close by. We wound up in the tiny hamlet of Garretson, South Dakota.

Garretson has two major claims to fame. For one, it's located near the red quartzite spires of Palisades State Park, and for another, it's home to the renowned Devil's Gulch.

Devil's Gulch, in case you've forgotten your Old West gangster history, is the spot where in 1876 the notorious outlaw Jesse James spurred his mount across a yawning rock chasm to escape a pursuing posse. It must have been the highlight of the week for Jesse, having botched the infamous Northfield, Minnesota, bank robbery a few days earlier. Two of his gang members were killed in the ensuing gun battle and two others were wounded, including Jesse's brother, Frank.

You can't help but learn these educational things when you get the opportunity to travel, especially when you visit a town so proud of its history that it slaps up numerous signs depicting a desperado on his horse vaulting across a gulch.

We decided to walk the footpath that loops around Devil's Gulch. It was quite pleasant, except for the fact that Chris and his pal had to stop on occasion, go to the edge of the rocky precipice, and chuck pebbles into the abyss. The fact that there weren't any guardrails didn't bother them in the least, but it sure was tough on MY nerves.

We finally found the fabled spot where Jesse James had jumped his horse across the gorge. A small bridge had been erected on the site to accommodate we non-horse-jumping pedestrians. The boys had to pause halfway across the bridge to gawk at the creek that flowed lazily fifty feet below. Downstream boaters on Split Rock Creek probably noticed large numbers of "loogies" floating by that day.

After we'd had enough of Garretson, we decided to return home via a different route and found ourselves meandering through a nondescript small town. My wife was driving (I always let her take the wheel; she can't read maps and my peripheral vision isn't all that good, so everybody wins), and I was staring blankly ahead. Without warning, an arm shot across my field of view.

"Look at that!" she exclaimed. I swiveled my head to see what warranted such urgent attention. Her tone of voice had caused me to believe that she had just espied a horseback posse that was galloping alongside us and trading gunfire with a pair of fleeing bandits. But all I saw were houses.

"Lookit that adorable Victorian! See that cute bay and the fancy turret?"

"Watch the road, Ms. Stewart!" I ordered as the car swerved. "You're gonna get nailed for DUIA: Driving Under the Influence of Architecture!"

My wife and I then got into a heated discussion about safety versus scenery. No sooner had the dust settled than the backseat boys got squirrely and we had to threaten them with several child welfare violations. By the time we got home, everyone was tired and crabby and I was actually looking forward to spending some quality time with the cows.

We continue to polish our day-tripping skills whenever possible. Because as Dad said that long-ago spring day on the headland of the cornfield, "You don't get twice as much work from a horse by working him twice as hard." ⬤

My Shameful Affair with the Farm Program

P lease, you must understand. Not necessarily forgive, but at least understand. After all, I am only human. I am just a man.

The whole affair began innocently enough. I was a young farmer, just married, and my life may have appeared idyllic. But something was missing. Something was gnawing at the core of my happiness. I'm ashamed to admit it: I could not produce a profit.

One day, when I went to the village for supplies, I ran into an old friend. We talked, and I shared with him my terrible secret. "Well," he asked, "aren't you enrolled in the Farm Program?" I told him no. He was surprised. "Everyone I know is in it. You'd better go check it out."

And so it was that I was introduced to that evil seductress they call the Program.

At first, everything seemed too good to be true. The

Program promised me much and asked but little in return. All she required was a small sacrifice: a tiny portion of land I must pledge to her. Idling perfectly good farmland is contrary to every fiber of a farmer's being, but I quickly acquiesced. I was flush with excitement and eager to please my new paramour.

I knew all about the Program's sordid past. I knew that she bestowed her favors on hundreds—nay, thousands—of other farmers. I also knew that her "business partner" was the all-knowing, all-seeing government. But I told myself that I didn't care. I pleaded to my wife that it was only for a season. Just this once, and then I would leave the Program. I swear!

During that early time, I would lie awake nights, wrestling with unanswerable questions. What is this "deficiency payment"? Was it in fact a form of welfare? Had I become a kept man? By the next winter, I had made up my mind: no more Program! But then, out of the blue, she sent me a love letter, in an envelope with a cellophane window. There was a generous check inside.

I was embarrassed by this overt flattery and uncomfortable about the check. But I found that I was able to cash the check despite my squeamishness. I began to see why the Program had so many admirers.

Each winter, I swore to my wife that this was it, that this was the last year with the Program. Ah, but I underestimated the wiles of the Program. Sometime after Christmas, the mail would bring a leaflet describing, in tantalizing detail, all the new benefits being offered by the Program. I would hide the leaflet in the bathroom, poring over it for hours and hours like a kid with a Victoria's Secret catalog.

In the words of the flier, I could hear her siren song, so soft, yet oh, so sweet. It was more than any mortal man could bear. I would feverishly punch the numbers through my calculator, and the last shreds of my resolve would evaporate.

When my wife was safely out of earshot, I would place a call to arrange a rendezvous. I would embrace the Program yet again, thrilling as she whispered words that sent shivers down my spine and into the very core of my checkbook. Phrases like "advance deficiency payment" and "nonrecourse marketing assistance loan" would put me into a swoon. And when it was over, we would bask in the afterglow and make promises to each other. In writing, of course. I would sign the triplicate forms and be on my way.

Upon returning home, guilt would overtake me and I would try to hide the evidence from my wife. She would see the ink stains on my hands and rifle through my sock drawer until she found the contracts that I had just signed. "You promised!" she would cry. "You said that last year was the last year and that you would give up the Program! I can't believe you anymore!"

And that's how it went, year after year, until a decade or more slipped by. The Program and I had a curious relationship. I recall one year when she asked me to not farm a third of my base acres. Preposterous! But as always, she wooed and enticed me. She offered me something called PIK certificates, which she said could be sold for cash or used to redeem grain that had been surrendered to the government under its commodity loan program. (The government doesn't farm, so I assumed the grain was ill-gotten. I knew better than to ask.)

That autumn, I was introduced to a kinky new move called "PIK and roll." I won't describe it here; suffice it to say that I am not proud of that period of my life.

Sometime later, I received a letter that was different from any other she had sent. No check, no comforting message, but a demand: She wanted money. Specifically, she wanted me to return part of my advance deficiency payment. I went to her representatives and pleaded that I had no such monies, that they had been long spent. No problem, they said. They would simply subtract the funds from my future payments. That's the Program for you. Always a class act.

Then came word that Congress, the Program's god-father, passed the sentence of death to my cherished paramour. As a punishment for her misdeeds, she would be forced to endure a lingering demise over a period of no less than seven years. Who but Congress could be so callous, so cruel?

But Congress argues that it's setting me free. I will be free, they say, to plant whatever and wherever I want, driven only by market forces. I can farm as God intended, unfettered by the whims of a capricious and half-witted Program.

And that is true. But what keeps me awake nights now is this: I will also be free to suffer the slings and arrows of a fickle market and perhaps go bankrupt without my beloved Program. Will I stick with the Program these next few years? Yes, yes, a thousand times yes. How can I abandon her at a time like this? These precious years will be our last hurrah, our final fling.

Perhaps many years hence, my wife and I will sit alone on a quiet evening. I will get that far-off look in my eye, and a single tear may roll down my cheek. I will be recalling

that glorious day when I redeemed all those PIK certificates and spent endless hours frolicking in that truckload
of wheat. Never was there such rapture, such profit. Those
were the days.

And my wife will know. She will know I am thinking
of the Program and, as the song goes, "looking back and
longing for the freedom of my chains and lying in [her] lov-
ing arms again."

Out in the Trees

I attended a family get-together recently. It was the kind of gathering where you sit around all day and drink coffee and chain-narf goodies while reminiscing about "the good old days." Can you imagine my shock and surprise when somebody at the gathering had the audacity to suggest that we were poor back when we were kids? After all, we had trees!

When I was growing up on our dairy farm, our trees came to mean more to me than mere protection from the elements. They were the currency of my imagination, the lifeblood of my entertainment world.

When pioneering homesteaders first arrived in this region more than a century ago, they found a prairie landscape that was as flat and barren as a stovetop. The homesteaders soon realized that our relentless prairie winds were more than just an annoyance; in the wintertime, a hard wind could be downright deadly. The pioneers began planting rows of trees around their homesteads, carrying buckets of water to the precious saplings during droughts, hopeful that the trees might begin to provide a windbreak within a decade or two. These shelterbelts, as they came to be known, are a living legacy that has been handed down to us from the pioneers.

Trees can work miracles when the TV is on the fritz and your parents are about to go bonkers because you and your brothers are wrecking the house as you perform a home version of *All-Star Wrestling*.

On such occasions, our parents would impolitely usher my two younger brothers and me out of the house and issue a stern edict: "Go outside and play awhile."

"Sheesh!" I would say to my brothers. "Bust one little light fixture and they give you the old heave-ho. C'mon, let's go out to the trees."

We often felt sorry for kids who lived in town. Sure, they might have some trees on their lot, but just a handful or so. And yes, there were trees in the park, but they had to be shared with all the other kids in the neighborhood. How could you build a decent secret tree fort if a couple dozen other kids knew all about it?

Thanks to the foresight of our farm's homesteaders, we had trees to spare. We had so many that each of the eight kids in our family could claim a tree as his or her own personal favorite and would still have an embarrassment of riches left over. To a young boy with a vivid imagination, this bounty was the true measure of wealth.

We would walk into our farm's shelterbelt and be instantly transformed. Sometimes we were fearless hunters, tracking tigers and lions through an unmapped jungle. Other times, we were the last of a contingent of French Foreign Legionnaires, locked in a desperate battle with bloodthirsty invaders who were intent on taking our fort. My favorite game, though, was playing Pirates.

I was always the Pirate Captain, and my brothers were co–First Mates. (Being the oldest had its advantages. To keep the peace, I secretly promised each of them that if the other were killed in action, he would become full First Mate.)

"Aarrg, ye scurvy dogs!" I would bellow. "Weigh anchor! Hoist the mainsail! Watch that jib! Hard to port! Steady, lads, or we'll run aground! Aarrg!" Talking like a pirate captain was the best part of being Pirate Captain.

Inevitably, we would be struck by a storm of such fury that a hurricane would seem like a tempest in a bathtub by comparison. Despite my heroic efforts at piloting our vessel, we would become shipwrecked and be marooned on a remote, uncharted island.

As we hacked our way through the island's steamy jungle, the crew might begin to grumble about the lack of rations. Just when I feared that mutiny could be imminent, we would stumble upon a small clearing where sat a farmer's hovel. Outside the farmhouse, a woman was hanging clothes on a line. What luck!

The crew might speak excitedly of plundering the farmhouse for silver and gold. I would tell them, "Nay, lads, these simple folk would have no such treasure. But see yon wench? What say we take her captive and make her do the one thing that all pirates dream of during those long months at sea!"

The crew gasped. "You mean . . . ?"

"That's right, laddies! We'll force her to whip up a batch of Rice Krispies Treats! Are ye with me? Then draw your cutlasses and . . . charge!!"

We burst from the jungle and quickly surrounded our surprised mother. "Aarrg! We be the wickedest pirates what ever sailed the seven seas! Get out the marshmallows or ye'll walk the plank!"

Mom would pause from her clothes hanging. "Put down those sticks before you hurt somebody!" she would admonish. "And go inside and wash your hands. How could you boys get so dirty in such a short time?"

Being a Pirate Captain wasn't easy. This was especially true when your mom made you peel potatoes as punishment for breaking a light fixture.

Somewhere out in our shelterbelt, we kids indeed found treasure, although it wasn't anything that could be measured in dollars and cents. It was a deepened appreciation for the wealth that's contained in all our imaginations. And also for the bequest left to us by our windblown pioneers. ●

Of Silos and Learjets

~~~~~~~~~~~~~~~~~~~~~~~~

I t's an onerous job, climbing the silo to set the silo unloader back up. Powdery, itchy grain dust tumbles off each rung of the ladder and swirls around inside the chute, coating the skin, getting into every orifice. The wind is never in the right direction on the days when you have to climb the silo.

This day is no exception and I climb the ladder with my eyes closed most of the way. At the top, I turn a latch, and the door opens inward, exposing a heap of golden corn. I crawl into the silo and, now safe, peer down the chute. Fifty feet is a long way if a guy were to lose his grip and fall. The gloomy metal chute is a vertical train tunnel; a bright patch of sunlight at the bottom is the locomotive's headlamp.

I set to work, sinking knee-deep into the loose grain as I put the silo unloader back into its operating position. The cavernous silo roof catches sound and amplifies it: Far off in the distance, I can hear the receding drone of a small airplane. And I think about what I had just heard on the news.

At that very moment, five miles up, a Learjet was streaking through the South Dakota sky on a journey that had a terrible, inevitable end. The crew appeared incapacitated, the broadcaster had intoned, and all anyone could do was watch as time and fuel ran out. Pro golfer Payne Stewart was among those hurtling through the sky aboard the doomed Learjet.

Five miles is a long way to fall.

I climb back down the chute and take off my shoes to shake out the corn. I know that each glistening yellow kernel represents both life and death: Moisten these seeds and they will sprout, bringing forth new life. A life that was made possible only through the death of the parent plant. I flip a switch, sending power up to the silo unloader's electric motor. The unloader sounds strangely distant up there at its highest altitude.

I give the unloader's winch a few experimental cranks, spooling out a tiny bit of cable. The unloader emits a hollow moan. No corn comes down the chute for some minutes. I know that the machine is scribing circles in the silo and hasn't yet made contact with the pile of grain. Mindless motion, not unlike that of a jet flying on autopilot. I turn the winch's crank and spool off some more cable.

Finally, I hear a few tentative kernels pinging against the galvanized steel chute, like the first drops of a summer-time rainstorm hitting a tin roof. Soon there is a surge of corn, a torrent that arrives upon frothy whitecaps of grain dust.

And like a thunderstorm, the flow abates almost as quickly as it started. I give the winch a few more cranks and look at the spool of cable. I can tell roughly how much corn is left by the number of wraps left on the spool. I will watch these wraps carefully, as they represent the amount of time remaining before I run out of feed. Hopefully, it won't be until next fall, when a new crop of corn will be ready to go into the silo.

I look at the corn lying at the bottom of the chute and idly wonder how many kernels are in the silo. I'm sure it could be figured out, but I really don't care to think about it. The ag economists say that this grain probably cost me more to raise than what I could buy it for—yet another thing I really don't care to think about.

I glance up at the pleasant sky and check my watch. By now, according to the newscast, the Learjet will have run out of fuel and fallen to the earth. Six people, we will learn by day's end, will have lost their lives in the incident.

Sweet dreams and flying machines.

I shut the silo unloader off and head for the house. And I muse to myself. I wonder if that golf pro woke up this morning and looked at his spool. I wonder if he smiled to himself, secure in the knowledge that he had a ton of cable left. "Go ahead!" he thought. "Give 'er a few more cranks!"

I guess nobody ever knows exactly when they'll run out of cable. And in the end, it really doesn't matter if you were up fifty feet or five miles. ●

What's in
a Cow's Name?

~~~~~~~~~~~~~~~~~~~~~~~~~~~~~~

"**C**ows may come and cows may go, but the bull goes on forever."

That old adage may ring true for some, especially in view of the lengthy sessions held at local coffee shops and feed stores. But any cattle person worth his or her salt block knows that in the real world, the exact opposite is the case: Bulls come and go, but cows are the bedrock of bovine society. In fact, a cow may last longer than a marriage nowadays, which means that the relationship between a farmer and a cow could qualify as "long-term."

As such, cows are often named. Some of these names are none too flattering. When I was a kid, Dad, who was the official cow-namer on our farm, bestowed such monikers as "Camel" (her face looked just like a dromedary's), "Ma's Cow" (which came out sounding as "Moscow"), and "That Durned White Biddy" (pretty self-explanatory).

But perhaps the most memorable cow Dad named was the one that he called C.C.

A name such as "C.C." is a pretty strange moniker for a Holstein cow. But it rolled off the tongue a lot easier than "Differently Abled Domesticated Mature Female Bovine."

C.C. started out as just another Holstein heifer in our herd. This all changed when she gave birth to her first calf, a difficult delivery that required massive amounts of assistance. The ordeal left C.C. partially paralyzed.

The outlook for a "downer" cow is generally grim. Future employment prospects are limited to such things as working at McDonald's—and I don't mean as a cashier.

But Dad must have sensed that there was something different about C.C. He resolved to do whatever it took to save her. I guess you could say he had compassion for the cow and also wanted to conserve her. So it was that we kids, who got the job of carrying feed and water to C.C., were exposed early on to that much-ballyhooed "compassionate conservatism."

C.C. thrived on all the attention. She ate like a horse and slowly but surely recovered her strength, eventually regaining her ability to walk. It was a glad day when she rejoined the herd, mostly because we were sick and tired of providing room service for a cow.

We pretty much forgot about C.C., but she obviously never forgot about us. When C.C. calved again, she was put into our stanchion barn to be milked. That is when her fondness for people came to light.

Whenever you bent over to prep C.C.'s udder, you would suddenly experience a rough, scrubbing sensation on your rump as C.C. enthusiastically licked your heinie. Such was her gusto that she would nearly hoist you off the ground with her tongue. She also had a penchant for nosing your shirt up and licking the bare skin beneath. It felt as though you were being scoured with slobbery sandpaper.

C.C.'s outstanding people skills didn't end there. If a person were to take a stroll across the cattle yard, C.C. would trot over to provide an escort. And if a dog was nearby, he'd best watch out: C.C. *hated* dogs.

I discovered C.C.'s animosity toward canines one spring morning when I was herding our milk cows into our stanchion barn. Our farm dog was hanging around nearby, superintending things with a vacant expression on his face. Without warning, a thousand-pound black-and-white missile shot from the herd and streaked toward the dog. The canine wisely beat a swift retreat. Yelping and whining, he leaped through the fence to safety. C.C., satisfied with this outcome, trotted over to me and began to lick my shoulder. She wouldn't leave even after I physically tried to shove her away. From then on, she would only enter the barn if she was walking beside a human. C.C. no doubt believed that she needed to "protect" us from that despicable dog.

I took advantage of this knowledge once when a buddy came over for a visit. He and I and our farm mutt were walking out to the field when I innocently suggested that we take a shortcut across the cattle yard.

We were tiptoeing through the cow pies when C.C. spied us. She came galloping down the hill at such a speed that my buddy judged us to be in mortal danger. He strongly urged that we get while the getting was good.

"Naw," I said. "Let's just stand here a minute."

C.C. thundered toward us. At the last possible milli-second, she swerved for the dog, who instinctively hightailed it for the fence. The victorious C.C. then trotted back and stood close by and mooed at us with motherly affection.

"Wow!" said my buddy. "She's quite a cow!"

"That's nothing," I said. "Bend over beside her like you're gonna milk her. And hike up your shirt just a bit."

Among all cows I have known, C.C. was by far the most human. As a downer, she shouldn't have survived for more than a few days, but during her many years in our milking herd, C.C. repaid our kindness by being a shining example of gratitude and affection. And also by scrubbing us as often as possible with the slobbery tongue of motherly devotion.

A Dog Named Sam

A horde of growling bulldozers and dinosaur-like excavators have carved streets into the farmland that once was the east half of Section 2, Medary Township, Brookings County—the farm otherwise known as the Old Revell Place.

What would be my interest in this example of inexorable urbanization? None, really, except that that was the farmstead where, many years ago, I launched my farming career.

As I drove past the Old Revell Place recently, I mused how ironic it is that developers choose names that describe the very things they destroy. I guess labels like "Pheasant's Nest" or "Prairie Vista" make for easier selling.

And I also wondered. I wondered if I should tell the developer how the soil next to the draw consists of a foot of black dirt, beneath which is a layer of "sugar sand" that can swallow a tractor or a combine in a single gulp. I wondered if I should tell them how that land down east tends to flood after heavy rains.

Needless to say, the Old Revell Place holds a lot of

memories for me. That rickety house that could never seem to keep out the chill winter winds. The old barn with its cobbled-together milking facilities. But perhaps some of the fondest memories I have are those of Sam.

Shortly after I moved into the Old Revell Place and acquired some dairy cows, I discovered one of the farm's most glaring faults: the fences. The combination of long-necked Holsteins and barbwire that belonged in a museum proved to be a constant headache. And so I hired Sam, my first blue heeler dog. He was nothing but a pudgy ball of grayish fur when he came to live with me. The pup and I soon became inseparable.

It wasn't long before Sam began to help me with milking. He would carefully watch the cows as they exited the milking parlor, sneaking up on them and nipping their heels just as they rounded the corner. I quickly found that owning a blue heeler was very similar to having another person with you at all times. Sam was always hanging around, watching whatever I was doing, seemingly worried that he might be missing out on something.

And he was a smart dog. For instance, I might be wrestling with a reluctant disk bearing when I would glance up and see Sam standing there, watching me. "What do you think, Sam?" I might ask, and—you may not believe this—Sam would run off and chase a rabbit! There could be no doubt that he was trying to metaphorically empathize with this futile struggle called "farming."

Sam grew up to be a supreme cattle dog. I had a big (2,200-pound) and mean Holstein bull who had learned how to fiddle with the gate until it became unhitched and swung open. He would then lead his harem on a foray across the farmstead, usually homing in on the garden so they could use their sharp hooves to trample our tender vegetable plants. Those who think that cattle can't think should watch a bunch of cows who are bent on maximizing their owner's irritation.

If I saw that the bull was working on the gate, I would call for Sam, kneel by his side, and whisper, "See that bull, Sam? Go get 'im!" Sam *loved* this. He would creep down to the barn, sneak around the corner, and nip the bull on the nose. The bull would turn tail and run; Sam would give chase, bloodying the bull's heels. Sam and I both enjoyed this immensely.

When our first son was born, we worried how Sam might react to this newcomer. We needn't have; Sam claimed the baby like he was one of the pack. When my wife went out to get the mail, she would often take along the baby in the stroller, and Sam would walk alongside as if he were the official guardian. Aside from his rabbit hunting, these were the only occasions when Sam would leave my side.

One fateful day, I discovered that my feeder calves had escaped from their pen. I called for Sam and commanded him to "Sic 'em!" I left the dog to attend to the calf-chasing and went to fetch my fence-repair tools.

When I got back, the calves were walking about nonchalantly instead of galloping for the safety of their pen. Where was the fifty-pound canine demon that should have been nipping at their heels? Strange. I called for Sam. Nothing. I walked down the fence line, and there he was. Sam was stretched out on the sod, motionless, looking as if he had been frozen in midstride. An unlucky kick from one of the calves must have broken his neck.

My wife and I buried him on the spot just as one might bury a fallen soldier on the field of battle. My son couldn't understand why we were doing this to his dog. My wife cried. I tried not to.

Tidy rows of houses have sprouted where I once grew rows of corn and beans. Someday soon, a bulldozer will wipe away all the remaining traces of those old fences that I struggled so mightily to maintain. Perhaps the dozer operator will spot a small smattering of bones and wonder whatever it was that might have died there. Or perhaps, decades hence, some suburbanite will till his garden and uncover a strange calcium deposit.

I wondered if I should tell the developers about these things. But then I thought: Nah. Let them find out for themselves. ●

Farm Corporate Jargon

~~~~~~~~~~~~~~~~~~~~~~~~~

Farm magazines just aren't the same as they used to be. In a bygone era, you could count on a farm magazine to supply some real "meat and potatoes" kind of literature. These days, reading a farm publication leaves you with the distinct impression that you just chowed down on a pile of rice cakes—a lot of filler, but not especially satisfying.

I don't blame the writers, or even the editors, of farm magazines for this sorry state of affairs. No, I place the guilt where it belongs: squarely on the shoulders of Corporate America.

In the old days, farm publications were largely mom-and-pop operations. The guy who was editor-in-chief was probably also the press operator and wore the hats of both main writer and field reporter. The lady who was in charge of writing the recipe page was probably also the proof-reader and head accountant.

The point is that these people didn't mince words when it came to the content of their writing. They would have called a hog house a hog house, never a "swine confinement facility."

Sadly, much of that has changed over the past fifty years as large corporations have wrapped their profit-seeking tentacles around one farm publication after another. As a result, farm magazines have assimilated large amounts of corporate jargon into their lexicon.

I wonder—what would things have been like if big corporations had controlled farm magazines a century ago, before farmers had radio and television to dilute the magazine's effects? This would have been a typical conversation between a farmer and his son:

"Okay, son, this morning I would like you to integrate the equine power units with the solid waste disposal apparatus."

"You mean you want me to hitch the horses to the manure spreader? Why?"

"Because you have been tasked with the oversight of the Solid Waste Management Team at the Juvenile Bovine Detention Facility."

"What?! I gotta clean the calf pens AGAIN? It's Wally's turn this week!"

"Your brother has opted to take a day of personal leave in accordance with stipulations in his employment contract."

"That slacker! I'll get him for this!"

"Please. Any interemployee disputes should be referred to the Head of Employee Grievance Services at the Human Resources Department."

"You mean Mom?"

"Um ... Yes."

"Can't you help with the calf pen, Dad?"

"Sorry, but my schedule is already full. Due to the restructuring of the Dairy Products Production and Procurement Division, Alice's position is being eliminated."

"What? You can't sell her! Alice is my favorite cow!"

"It's a done deal. The Board of Directors decided—"

"You mean you and Mom?"

"Yes. The Board determined that Alice has consistently failed to meet production standards and has been noncompliant with corporate policy of utilizing synergism as a modality for achieving the company's objective of increasing employee cooperation."

"You mean you're gonna sell her because she's ornery and doesn't give much milk?"

"Yes. It is regretful, but we must do what we can to fulfill our Mission Statement, which adopted the goal of increasing our equity position. We gave Alice a bon voyage party last night, and the CEO—"

"You mean Mom?"

"Yes. The CEO gave a moving speech praising Alice for her years of faithful service and, at the end, presented Alice with a handsome Certificate of Appreciation that is suitable for framing. I also made a small speech and presented Alice with an ear of corn as a token of my esteem. Alice seemed to enjoy the festivities . . . Well, she liked the ear of corn anyway. I'm surprised you weren't there, son. I sent you a memo."

"If I read all your memos, I wouldn't have time for anything else." The boy wipes a tear from his eye. "I'm gonna miss old Alice."

"There, there, son. If you like, you could visit our Employee Psychological Counseling Representative at the Human Resources Department—"

"You mean Mom?"

"Yes. But first, be sure to fill out Form HR 876—in triplicate, of course. One copy goes to Human Resources, one you keep, and one will go into your permanent file at the Bureau of Employee Records."

"Naw, I think I'll pass. All I really need is one of Mom's chocolate chip cookies."

"Fine, but be sure to remove all hydrokinetically adhered organic and inorganic compounds from your footwear before entering the residential facility. I don't want any more acrimonious memos from the Assistant Supervisor of Janitorial Services."

"You mean sis gets mad about me tracking in mud?"

"Yes. By the way, Dilbert, what's the anticipated time line for moving forward with that calf pen project?"

"I'll have to get back to you about that situation, Dad. I'll be sure to send you a memo."

A Lesson in Organic Chemistry

~~~~~~~~~~~~~~~~~~~~~~~~~~~~~

I t was a hot summer Sunday and I was rushing through morning chores at the hundred-cow dairy my wife and I operated in partnership with my parents. I had made plans to meet my wife and our two young sons at the arts festival in Brookings and didn't want to be late.

While scraping the barn alleys with the skid loader, I saw that the manure pump had plugged. This would mean an annoying delay. Our cows were free to wander the alleyways of our dairy barn, and we used our skid loader, a smallish tractor-like machine, to daily push their dung into an underground manure pit. A tractor-powered pump was then used to move the manure to a nearby earthen basin for longer-term storage.

There were two ways to unplug the pump. The right way involved hoisting the pump from the manure pit with a front end loader, but that would cost a huge chunk of time. Then there was the quick and dirty method, which involved climbing down into the manure pit and using a spud bar to clear the slug. I chose the quick and dirty route.

I had descended into the pit and begun to slam out the slug when I felt woozy. "It's the gas!" I thought and began hurrying out of the pit.

I nearly made it. I was almost to the top; I could see the blue sky and could hear our John Deere 4020 tractor idling. Then the world faded to black.

The pit had contained hydrogen sulfide gas, also known as H_2S. Hydrogen sulfide combines with oxygen in the lungs to become H_2SO_4, or sulfuric acid. This powerful corrosive strips away the lining of the lungs, causing victims of H_2S to drown in their own bodily fluids.

My parents wondered why I hadn't yet left the barn. They investigated and found me floating faceup in the manure pit. The first responders were summoned, and my unconscious body was hauled from the pit. No one could find my pulse. I wasn't breathing. An ambulance soon arrived, and its crew immediately began to administer CPR. The EMTs intubated me and used a bag to make me breathe.

I was rushed to a local hospital. One of the first responders was eventually able to locate my wife and alerted her to the situation.

The doctor who worked on me in the ER told my family that my condition was extremely serious and that I would likely pass away soon. My wife found that unacceptable and demanded that I be transferred to a larger hospital. After she was told in no uncertain terms that there was no hope, her request was fulfilled.

A team of doctors worked on me at the larger hospital. After I was stabilized, my family was informed that I had a fifty-fifty chance of survival—if I made it through the next week.

At the end of my first week in the hospital, it suddenly became almost impossible for me to breathe. This despite the fact that I was still intubated and was hooked up to a state-of-the-art ventilator. Try as it might, the ventilator could no longer push air into my lungs. It was theorized that my lungs were swelling due to their severe injury and that nothing more could be done. My wife was told to call in the family to say their final good-byes.

She instead asked the physician to consult with the Mayo Clinic. The doctors there suggested an endoscopic examination of my lungs. This was done, and it was found that blood clots were plugging my bronchial tubes. The clots were removed. I could breathe again.

My condition improved slowly but steadily and I

was able to walk out of the hospital five weeks after being helicoptered in. The only residual effect is a reduction of my peripheral vision due to the death of some cells in my brain's vision center.

I think about my accident every day. I'm reminded of it every time I swallow and feel the tug of my tracheotomy scar.

I am grateful for each new sunrise. I know full well that I would be pushing daisies were it not for my dear wife's stubbornness and the help of many good people.

A few days ago, I was digging in the cupboard, looking for something or other, when I ran across an item that caused me to totally forget what I was looking for.

The artifact was an old medical report, a paper monument to my manure pit accident. The records document my long and complicated process of recovery, much of which I don't recall thanks to anoxia and large doses of narcotics. I received so much morphine during my month in intensive care that I became addicted.

The few and faulty memories I have of my hospital stay are the reason my wife obtained the copy of my medical records. Otherwise, I might deny that it happened and claim that she and my family had invented it all as part of some kind of mind-messing conspiracy.

One of the first items that jumps out from the front page of my medical report is the word "Pavulonized." Pavulon is a drug that is used to totally immobilize a person. Which figures, as the word "combative" appears in the same sentence. Apparently, I was a very bad boy.

Another term that is prevalent throughout the report is ARDS, which is the abbreviation for Adult Respiratory Distress Syndrome. Inhaling hydrogen sulfide and manure has a tendency to earn a person that particular acronym. Survival is iffy at best, as proven by the attending physician ending the first page with "Prognosis is guarded at this time."

My medical report is shot through with a hodgepodge of weird, unpronounceable medical jargon. Words like "encephalopathy" (they were worried about brain damage) and "febrile" (I had a fever) and "pneumothorax" (my right lung had collapsed).

The report also contains excruciating details regarding my tracheotomy operation. If the sentence "The thyroid gland was mobilized upward somewhat and a tracheal elevator was placed in the second ring to elevate the trachea" doesn't give you the willies, nothing will.

Another fun-sounding procedure was the installation of a Swan-Ganz catheter: "Using the Seldinger technique, the arrow introducer kit was placed without difficulty in the right subclavian vein."

Other assorted terms that leap out from the pages: Radionuclide. Staph aureus. Diffuse nodular infiltrates. Acute pulmonary failure. Cerebral ischemia.

The report paints a picture of someone who escaped death by the skin of his teeth. It is a monument to the miracles of modern medicine and to my wife, who kept her wits about her and kept after the doctors to ensure that every one of those miracles was made available to me. It also paints a portrait of a guy who should be grateful for every single day he spends walking around aboveground. Which I most certainly am. As an old farmer neighbor of mine remarked, "You're the only guy I know who could fall into a hole filled with crap and come out smelling like a rose!" I couldn't argue with that.

You Stinker!

T he moniker "stinker" has been applied to me for nearly as long as I can remember.

It started out as a term of endearment. Back when I was but a wee lad, phrases such as "What has that little stinker gotten himself into now?" and "How did that stinker manage to sneak a baby pig into the house?" were often uttered with an air of amused affection.

As time marched on this gradually changed. Gone was the tone of amusement, replaced by the distinct aroma of annoyance.

"Why do you have to be such a stinker?" was pretty much shouted at me one day when I was twelve and it was discovered that my little sisters' dolls had been kidnapped. A cut-and-pasted ransom note demanding that these sisters' desserts be donated to me in order to secure the dolls' safe release was ascribed to me. Unjustly so, I might add. As I pointed out at the time, all the so-called evidence was circumstantial.

By the time I entered high school, "stinker" had taken on a decidedly sinister connotation. I frequently heard the word hooted behind me as I walked down the hall between classes. I often wondered who they were talking about and felt sorry for the poor slob, whoever he or she may have been.

It sure couldn't have been me. It's true, sharing a single bathroom with seven siblings made it impossible to bathe every day. But I always cleaned myself as best I could during the few minutes between the end of milking and the arrival of the school bus.

A few years after I began dairy farming on my own, I was fortunate enough to acquire a wife. Unfortunately, she was a city girl and hence unaccustomed to the earthy fragrances associated with bovine waste emissions.

She would often wrinkle her nose at the aromas I dragged into the house on my coveralls. "Don't you smell that?" she would exclaim.

Taking a deep breath, I would reply, "I sure do. Smells like money!"

My wife frequently hinted that such odors didn't belong in our house. She suggested that I install a laundry, and perhaps even a shower, down in the barn. And while I was at it, I could just as well add sleeping and cooking facilities. There would thus be no need for me to ever leave the barn, she said, and she could visit me on special occasions, such as Election Day and Leap Year Day.

My dairy farming days are now behind me, but I again recently found myself struggling with a smelly situation.

The car I drive for work purposes was previously used by a dog owner. This in and of itself is not an issue. The problem is that this person thought so much of his or her pooch that he or she couldn't bear to leave the dog at home. Fido must have ridden along on a regular basis, because I am exposed to a very potent dose of doggie odor whenever I drive. I like dogs as much as the next person; the problem is, the next person might not like dogs at all, or might even be allergic.

I have tried numerous fixes for this problem. I have cleaned the carpets and the seats, but to no avail. I have even purchased those tractor cab freshener pouches, assuming that a product that can eradicate the aroma of diesel smoke ought to be able to handle a little dog stink. No dice.

My wife advised me to try a particular odor-eliminating product. I won't say what brand it is, but will tell you that its name sounds a lot like "fur breeze." Which suits me just fine, since I'm sort of a furry guy, who enjoys the feel of wind in my, um, hair.

So I tried that fur breeze substance and finally succeeded in stifling the stench of dog in the car. I found, however, that in order to fully do so, I also had to spray some fur breeze into the air vent inlet. I have no idea how the car's previous driver managed to get his or her dog up into the car's air vent inlet, nor do I want to know.

If only I'd learned about this fur breeze stuff back when I was a newlywed! I bet I could have cut back to maybe one bath per month.

There is a group of people who are disdained for the way they smell. Not so much for the methods they use to smell, but more for the aromas they exude.

This downtrodden demographic is commonly known as "guys." The subgroup of this demographic that suffers the most is known as "husbands."

Let's face it: Women have a much more acute sense of smell than men. Females can detect a single malodorous molecule at a hundred paces. Guys, on the other hand, can't tell if a baby needs changing until the diaper's odor has become strong enough to stop a charging water buffalo. This is just one reason why women are generally better mothers than men.

It's nearly always news to a guy when his wife informs him that he stinks.

Skillful husbands can intuit when their impending personal odors might be offensive to the female olfactory system. As such, many husbands have perfected the tactic that is known universally as "silent but deadly."

"Oh my Lord!" a wife may suddenly exclaim in the middle of a gripping episode of *Better Call Saul*. "What did you eat? Week-old roadkill?"

"What?" the husband might protest. "I don't smell anything. It was the dog."

"We don't have a dog."

This is why the marriages that last the longest are those that involve household pets.

~~~~~~~~~~~~~~~

F oot odor has long been my personal albatross. Taking my clodhoppers off at the end of a long, hot day could empty the house. It was often broadly hinted that my socks would qualify as an EPA Superfund site. I, on the other hand, didn't think that my feet smelled all that bad. But then again, that's also how I feel about lutefisk, a traditional Scandinavian treat that's made by steeping cod in lye. This gives the fish a particular odor that has elicited responses from my wife that have included "Oh my God, what reeks?" and "I think I'm going to be sick!" Which just proves that some people have no taste.

Once, when I was a teenager, I acquired a blazing case

of athlete's foot. The simple act of removing my boots and peeling away my socks was pure agony.

A search of the medicine cabinet yielded a small bottle of a liquid athlete's foot remedy. Ignorant regarding such things, I doused my toes with a generous squirt of the stuff. I was later told that my bellowing could be heard several miles away.

I shared this tale of woe with an uncle, who replied that he obtained a wicked form of athlete's foot while serving in the Merchant Marines. Upon returning home to the farm, he was able to cure the malady only after he cut the toes off his boots, thus exposing his tootsies to fresh air and sunshine. He was a tough guy, so I wouldn't be at all surprised if he had done this in the dead of winter.

Then there is the issue of BO. This can be a tricky topic, as there are no objective standards regarding what passes for "a little funky" and what constitutes "a stink that would scour the rust off a plowshare."

For instance, I was recently yakking with an elderly bachelor dairy farmer when he asked if I could give him a lift to his pickup, which sat on the headland of a nearby field. I said of course.

He climbed into the car, and the air instantly filled with the choking aromas of dried dairy cow manure, diesel fumes, and old-guy BO. These odors intermingled with a level of bad breath that can only be obtained by washing down a cud of smokeless tobacco with a quart of stale coffee. I rolled my window down as we bumped our way to the field.

"Something wrong with your eyes?" asked the old guy.

"Must have got some dust in them," I replied as I gulped fresh air.

"Well, they sure are watering a lot. You should probably get that checked out."

"I've been meaning to ask," I said, between deep tokes of outside air, "what's up with that car air freshener hanging from the back of your cap?"

"That's to keep the skeeters away. Might look silly, but it works pretty darn good!"

"What about the smell?"

"That there's a bonus feature. Wearing a air freshener all the time cuts way down on body odors. I bet I take only half as many baths!"

It was hard to imagine how he's still single.

Experiments in
Fermentation

~~~~~~~~~~~~~~~~~~~~~~~~~~~~

There are farmstead wineries popping up all over the countryside these days. It seems that wine-making has become the "in" thing to do if you happen to own an old barn and a few acres of land. The land is for growing grapes; the barn is to give the winery "ambience." And the older and more run-down the barn, the higher the ambience.

I personally have nothing against these barn-based enterprises. In fact, I am actually quite familiar with the fine art of barn winemaking. I was an early bloomer, having brewed up my first batch of barn wine when I was but twelve.

That was the year when I got wind of this intriguing fungus known as "yeast." I had learned in science class that this seemingly innocuous microbe was the driving force behind a process called "fermentation" and was, by extension, responsible for such things as hangovers and many of the funnier scenes in the movie *Animal House*.

That summer I decided to test this process with some extracurricular scientific experiments. I enlisted the assistance of my two younger brothers, and after scrounging in the kitchen, we were able to assemble a winemaking kit. This included an empty quart-sized pop bottle, a can of frozen grape juice concentrate (purple, of course), a few cups of sugar, and a packet of dry yeast that we "borrowed" from the cupboard. And a balloon. You can't make wine without a balloon.

We hauled our wine makin's up to the hay mow, where, deep in the muffled sanctuary of the straw bales, we poured the thawed grape juice concentrate into the pop bottle. There was some debate over how much sugar to add, but it was finally decided that if a little was good, then a lot must be better. We added the purloined yeast, shook the bottle to mix the concoction, then capped it with the balloon.

Daily we checked on the progress of the pop bottle. As the yeast performed its mysterious alchemy, the balloon grew to an alarming size. It looked as if the quart bottle had sprouted a latex basketball. My brothers wanted to conduct a taste test almost immediately, but I said no, that we should guzzle no wine before its time. I told them that we would have to age it quite a while. At least a week.

A few days later, the three of us sat on a straw bale and, with great giddiness, unballooned our very first bottle of vintage barn wine. A fragrance slightly reminiscent of bread dough wafted throughout the hay mow.

I allowed my youngest brother the honor of the first swig. It's hard to describe the bittersweetness of that magical moment. The sweet part was watching my brother's expression as he took a big pull on the bottle, because the wine was definitely the bitter end of the deal. That stuff was caustic enough to cut through a plowshare.

I pretty much swore off of wine from then on. That is, until some years ago, when I took the family out to the West Coast to visit my wife's uncle and aunt.

Doris and Jim Granflatten lived in the midst of wine country, and it was they who suggested that we go on a wine tour. Actually, it turned out to be about the best part of the trip: The previous couple of days we had spent in LA, doing the Disneyland thing with a couple of tired and whiny kids, rubbernecking at the tall buildings, battling with the traffic, and generally behaving like tourists.

It was a pleasant change then, when we retreated to the (relatively) sparsely populated hinterlands, where the sky was once again blue, instead of yellowish-brown, and we no longer had to chew in order to breathe.

Central California boasts numerous wineries, some of which are situated by the roadside like lemonade stands. As you might imagine, wine-tasting tours are an extremely popular recreational activity, even though it's one that requires a designated driver.

The taste master at one of the first wineries we visited instructed me in the proper method for evaluating wine. "You must take your time and swirl, sniff, and sip," she said. "Don't gulp it down and say, 'Wow! That'll put hair on your chest!' And gargling before swallowing is a definite no-no."

After we had stopped at about half a dozen wineries and sampled two or three wines at each one, I was beginning to get the hang of things. It was quite fun, really, touring all those Old World–style buildings, savoring the balmy Mediterranean climate, and talking about how this wine has a fruity bouquet or how that one has an oaken finish. It was enough to make me feel tanned and sophisticated, unlike the pallid and clueless tourist from the hinterlands that I was.

Yet the ambience didn't seem quite right. "Say," I finally asked a winery employee, "something is missing here. You wouldn't have a straw bale for me to sit on, would you?"

Winter Storm Stories

~~~~~~~~~~~~~~~~~~~~~

PART I

I recently made the mistake of purchasing some clothing for my wife. I was quickly re-reminded that buying clothing for a woman is always a mistake, and that once the gift is made, the only thing that remains to be determined is the magnitude of your blunder.

If she reacts with words such as "I don't think this will work for me," what she really means is "You have no clue what size I wear, do you?"

If the gifted clothing elicits a response of "That's really not my style," what she might mean is "Why do you suddenly become color-blind whenever you go to the store?"

After all these years, I finally learned that there are only four things a husband should give his wife: money, chocolate, jewelry, or flowers. The ideal gift would be a box of chocolates that's topped by a bouquet of flowers that contains a necklace wrapped inside a hundred-dollar bill.

We managed to make it all these years despite my gifting cluelessness. It's safe to assume that we will remain married, due to the fact that my wife no longer introduces me as her first husband.

Also, when asked how long we've been married, she has quit saying, "I don't even want to think about it!"

My wife and I recently spent a three-day weekend at home together. It wasn't planned that way; the weather closed in and we became snowbound.

About halfway through the third day, I was informed of a calamitous development. According to my wife, the situation was as horrifying as a nuclear reactor meltdown or being forced to watch Carrot Top.

In fact, she claimed it was worse than either of those. We were clean out of toilet paper.

I suggested that there are alternatives to Charmin. Newsprint sprang to mind, followed closely by recycled plant material—specifically, the cobs I had stashed in the chicken coop for just such an emergency.

Both ideas were swiftly shot down.

As the Man of the House, I knew that there was only one thing to do: I had to brave the swirling sea of snowdrifts and voyage into town to requisition a new supply of Charmin.

It wasn't my first rodeo, so I knew what to do. I girded my loins with insulated coveralls and strapped on my heavy winter boots. I threw a snow shovel into the back of the pickup for that "worst case" scenario.

Tearing along on our township road, I blasted through the forward guard of drifts. Turning a corner, I halted, gob-smacked. Endless snowdrifts stretched out before me like an angry, frothing ocean.

What to do? The term "no guts, no glory" popped into my head. With the theme music from *The Good, the Bad and the Ugly* reverberating through my skull, I locked my pickup into four-wheel drive and floored it.

The pickup punched through drift after drift, until the drifts all became an unbroken wall of snow. The windshield was blasted with white powder, blinding me with whiteness and eliciting no small number of muttered imprecations.

Suddenly everything became quiet. The pickup was running, but all forward motion had ceased.

I was completely and utterly stuck.

The shame, the indignity! It's been years since I'd been that stuck. You might think I would have learned.

I initiated plan B, which consisted of clambering into the back of the pickup to retrieve the aforementioned snow shovel. Ten minutes of digging—the kind of frenetic shoveling that's often associated with a person who is in the grip of a full-blown panic attack—uncovered little more than the pickup's front bumper. I had forgotten what it's like to shovel out the snow from beneath an entire pickup and that it usually involves moving several tons of snow.

This left me no choice other than plan C, which involved hoping that my cell phone would be able to locate someone to rescue me. Hands trembling, I dialed the number of a kindly neighbor. A breath of relief burst from my lungs when he answered on the second ring.

The neighbor said he would gladly come over and pop me out with his pay loader, but that it would be half an hour or more before he'd get there. I thanked him profusely and hung up, knowing full well that this would be a "you'll never believe what that idiot did" story the next day at our local coffee shop.

There was nothing more I could do other than to sit and listen to the wind moan, and stew in my humiliation as I waited, helpless as a turtle atop a fence post.

Minutes later, the township's snowplow appeared and began to head toward me! Woo-hoo! Freedom!

The snowplow poked its gigantic steel nose through the drifts that held me in their icy grip. We hooked the snowplow onto my pickup and I was yanked out as slick as a loose baby tooth.

I chatted briefly with the snowplow driver, who was kind enough not to ask, "What sort of idiot would try to go through THAT?!"

Turns out he wasn't there to rescue me, but was simply plowing our roads. Had I just waited half an hour I would have never gotten stuck.

On the plus side, I'm sure that my reckless risk-taking furnished a good deal of fodder for the next morning's coffee klatch. It could thus be argued that my getting stuck was actually a form of public service. At least that's what I told myself. It helped me feel slightly less stupid about the whole affair. My excursion to the store ended our toilet paper emergency. And for once, the stuff I brought home from a shopping expedition was both the right color and the right size.

Which is outstanding, because I have now discovered an entire new category of acceptable gifts.

PART II

I n this part of the world, cold is a palpable substance, as real and as hard as an anvil. Those of us who have lived through the crucible of the cold are imbued with an attitude and a set of values that are particular to Northerners.

The attitude could best be summed up with the words "It could be worse." After enduring just one of our average winters, we know that no matter what manner of catastrophe might befall us, we should never complain, because it could always be worse.

This is because it has often been worse. At least that's the consensus of those who talk about the weather, a segment of the population that includes approximately everyone. If you live in this region and don't have a good story to tell regarding nasty winter weather, you may suffer from a lack of imagination. Don't let the misfortune of not having experienced a spate of bad winter weather stand in the way of a riveting tale.

When I was a kid, the talk at Christmastime gatherings inevitably turned to the weather. "This winter might be a tough one," one of the adults might intone, "but it isn't anywhere near as bad as the winter of forty-two." Heads would nod in silent agreement.

Someone else would say, "The squall we had last week was nothing compared to the blizzard of fifty-four. Now, there was a storm!" Heads would again bob in concurrence.

Not wanting to be left out, I would say, "That Christmas snowstorm of sixty-one was really something. The snow was belly deep."

Never mind that I had been only four years old in 1961 and that "belly deep" is a relative term. Never let the facts stand in the way of a compelling story.

The cold has also instilled us with such tidbits of wisdom as "Fogged eyeglasses do NOT indicate a heightened level of romantic feelings" and "If you can see your breath when you wake up, the furnace has probably run out of fuel" and "It's always the husband's fault when the furnace runs out of fuel."

Most of this is just common sense. But it's surprising how uncommon common sense can be.

For example, one would think that nobody should have to be told, "Shut the door, you're letting all the heat out of the house!" Yet you would not believe the number of times we had to say that to our boys when they were growing up.

Another ingrained value is to always Be Prepared. It's no coincidence that the founder of the Boy Scouts was a Northerner.

Being Prepared means having enough food on hand to survive a three-day blizzard. This is why the mere mention of snow flurries by the weatherman can cause a stampede of panicked "It's the end of the world!" shoppers to descend on the supermarket. My wife and I have sometimes been caught up in this groupthink, even though our cupboards already held enough food to feed a troop of Boy Scouts for six months. The Be Prepared instinct is that strong.

The other night, shortly after our most recent cold front roared through, I stepped outside for a moment. It had dropped to ten below under a clear sky; the stars were so bright and sharp, you could grab Orion by his belt.

And the stars were making noise. Specifically, they were producing a stream of honks and squawks. It took a moment to realize that a flock of geese was passing overhead through the black, brittle ether.

I don't speak goose, but the squawks seemed to be saying, "It was YOUR dunderheaded idea to migrate in December! I'm freezing my beak off! Why couldn't we be like the Gundersons? THEY don't wait until the last minute! THEY always head south at the end of October!"

Which reminded me that it was time to check on the furnace's fuel level. Because bad as this cold weather might be, I certainly didn't want things to get worse.

One November weekend, rain fell as the mercury plummeted to the freezing point. Ice formed on every available surface, including the card slots on ATMs. Frigid bank customers waited patiently as the person at the head of the line struggled to get the robotic teller to accept his or her card. This would be a bad storm.

As the ice built up, trees and power lines sagged and snapped. Then came the wind, a full-throated nor'wester that howled and screamed, bringing air that smelled like permafrost, a special delivery direct from Siberia.

We denizens of the prairie have a name for such conditions: the kind of weather that keeps out the riffraff.

We lost power at our place on the first morning of the storm. My wife had opted to skip work that day, reasoning that the closing of the interstate highway system was a sign that she should remain safely at home instead of gambling with the skating rink–like roads.

We assumed that the power would be out for only a short time, since this is usually the case. We reported the outage and waited. And waited. And waited some more.

I called the power company again and was told that power lines were snapping faster than castanets and that it could be a day or more before we had electricity again. The wind had kicked in by then and was blowing at speeds normally associated with the jet blast of an airliner. The house began to cool at an alarming rate and my wife began to fret about the possibility of frozen pipes.

I was secretly gladdened by this development. Having grown up on the prairie, I knew that one should always Be Prepared. In our case, it meant having a wood-burning furnace and a supply of seasoned firewood in the basement. As my wife lit candles, I descended into the bowels of our dusty cellar and coaxed the old wood furnace back to life. It was strangely thrilling and primal, knowing that so much depended on my fire-making skills.

Upstairs, I reported my triumph with the wood burner and was hailed as the Master of Fire and He Who Keeps the Pipes from Freezing. "Isn't this romantic?" asked my wife as she lit what appeared to be the hundredth candle. "It's just like *Little House on the Prairie*."

"Sort of," I agreed. "Except for Ma and Pa Ingalls didn't have a four-wheel-drive pickup at the ready in case they decided to run for groceries."

Things sure are different without electricity. There's no heat, no light, and, since we depend on a well for water, no H_2O.

As far as I was concerned, the lack of water was the

least of our problems. I was born into a farmhouse that had no indoor plumbing, so going without running water seemed to be just a minor inconvenience at worst. But like the males of many species, guys see the entire out of doors as one giant bathroom.

My wife and I generally don't mind spending a quiet evening at home. Normally, we watch a little TV, and I might surf the web or write a bit on the computer. With none of these things available, we had to resort to conversing with each other. It soon became quiet. Much too quiet.

"Do you have to do that?" she suddenly blurted.

"What? You mean eat this pickled herring?"

"Yes! That stuff reeks, and now so do you!"

"Hey! My Norse ancestors survived many a harsh winter on pickled herring! If it was good enough for them, it's good enough for me!"

Quiet again descended, accompanied by tension. To assuage the unease, I scrounged up a transistor radio and found some batteries. We were soon listening to a station that was broadcasting from San Antonio. "Cooler tomorrow with a high of fifty-eight," intoned the announcer.

"Fifty-eight!" lamented my wife. "What I wouldn't give to have it that warm again!"

The house was right at fifty-eight degrees when we

awoke the next morning; it seems that I forgot to get up during the night and stoke the furnace. "That does it, fish-breath!" said my wife through chattering teeth. "I can't take it anymore! You NEED a shower! My scented candles can only do so much! Either you get us some electricity, or I'm leaving you for the Super 8!"

Not knowing what else to do, I fired up the pickup and drove north, following the power line that feeds our house. A jolt of joy shot through me when I espied a lineman working on the line. I could have hugged him.

"Am I ever glad to see you!" I called out. "I think you just saved a marriage!"

Visiting

~~~~~~~~~~~~~~~~~~~~~~~~

There are things wrong with today's world that I believe can be traced back to customs we have abandoned. For instance, we were a much more civilized society when men wore double-breasted suits and snap brim fedoras. Why did we ever give them up? Nothing lends a guy a sense of savoir faire like a well-fitted suit and a matching hat. If your suit is pinstriped, you might be mistaken for a mobster, but that could actually be quite helpful when you're trying to get a good seat at a trendy restaurant.

And whatever happened to ladies wearing hats, especially the kind of headgear that looked as if it were the result of an accident involving a flower shop and a truckload of ostriches? Or even those elegantly understated pillbox toppers that were made popular by Jackie O? And when did the practice of ladies wearing elbow-length evening gloves fall out of favor? Why did ladies quit wearing those elegant and billowy skirts, the kind that contained more cloth than a parachute?

Our behavior would improve greatly if we went around dressed to the nines. The glow of our panache would light our way wherever we went.

Another custom that has fallen into disuse as of late is the fine art of visiting.

When I was a youngster, people would spontaneously drop in on one another for spontaneous visits. This could happen at any time, but most commonly occurred on Sunday afternoons. There was no such thing as calling ahead to make an appointment.

And no, this was not because telephones hadn't yet been invented. We had even moved past the string-and-can technology and were using actual phones. But the fact we had to use party lines made our phone calls about as private as a radio broadcast.

Back then, people simply expected visitors, or "company," as it was called. Very often our parents would pack us kids into the station wagon and go visit our grandparents.

At first blush, that sort of thing may have seemed like torture for a little boy. After all, my parents and grandparents tended to just sit around the living room, yakking and drinking coffee. What excitement is there in that?

But most of the time, some of my cousins and their

parents would also be visiting. This meant there were playmates available for games of tag or hide-and-seek. For some reason, they never even tried to find me whenever I hid. I guess I was just too good of a hider.

At about midafternoon, the call of "Lunch!" would echo across the farmyard. Lunch was something you ate at the midpoint of both the forenoon and the afternoon.

We would tear off for the house and gather around a table that brimmed with cookies and pies and other such yummies. None of these goodies contained so much as a single calorie, as stuffing ourselves with them never caused us to gain a single ounce. After filling our bellies with these treats, we would dash back outside to tear around some more.

Speaking of treats, it was customary for a caller to bring along something tasty. This was not just polite; it also served as an excuse: "I just baked this huge batch of chocolate chip cookies and don't know what to do with them all. Suppose you can help me eat them?" Dumb question.

But not having goodies with you didn't mean you couldn't pay a visit. Even empty-handed, you could still randomly stop at someone's house and could depend upon being invited in for "coffee and a bite." Even our neighborhood Norwegian bachelor farmers would extend this invitation, but you had to be careful there as their version of "a bite" might involve a lard sandwich.

After a visit, it was expected that the visitees would soon return the favor by calling upon the visitors. At our house, a supply of treats was kept on hand for such unexpected drop-ins. We kids constantly nibbled at these treats to keep tabs on their quality.

Our neighbors Al and Lorraine didn't do much visiting, but let it be known that they thoroughly enjoyed having guests drop by.

Lorraine was a top-notch cook who produced mass quantities of marvelous munchies. Their farmhouse was as busy as Grand Central Terminal.

Once when my wife and I stopped in, Lorraine complained about how portly their sheltie, Brownie, had become. "I even bought her this special diet dog food!" said Lorraine, opening a cupboard that contained enough canned pooch food to feed a herd of shelties.

I glanced at Al, who had his hand under the table. Brownie, who was also under the table, was quietly eating the cookie he held. Al grinned and winked. I held my tongue and grinned back at Al as Lorraine groused about the utter worthlessness of that fancy diet dog food.

It soon became time for us to leave. As we got up to go, we reminded Al and Lorraine that it was now their turn to visit us. They promised that they would try to stop by sometime.

I then plopped my snap brim fedora onto my head and set it at a jaunty angle and helped my wife slip into her elegant, elbow-length gloves.

The Four Seasons
of Farming

~~~~~~~~~~~~~~~~~~~~~~~~~~~~

When winter finally retreats for good and spring sweeps across the prairie, my farmer's heart beats just a little faster.

As the land warms, it gives off an earthy aroma that makes me think of bread dough rising in a sun-drenched kitchen. The breeze caresses my ear, whispering sweetly of unlimited possibilities. I soon develop an irrepressible itch to drive my tractor out into my field and till the earth. Planting crops in the luscious black soil gives me a deep sense of joy, similar to what I felt that gentle June evening more than thirty years ago when my wife took me in her arms and murmured, "We're going to have a baby."

But spring tarries only a short while and soon gives way to summer. This change can be abrupt; within the span of a week, I have gone from worrying about a late frost to fretting about the scorching heat.

Life enters a phase of growing, nurturing. The days wax longer as the summer solstice approaches. The sight of fat baby calves and their mothers luxuriating in the deep grass makes me smile. A father robin warbles from a treetop, filling the air with his song of joy. Down in the marsh, a mother Canada goose honks proudly as she glides across a mirror of water, her brood of chubby gray goslings in tow. Fireflies sparkle in the twilight like slow-motion shooting stars.

We farmers are at our busiest now, making the most of this balmy season. What fragrance better portrays summer than the scent of a field of freshly cut alfalfa? When they gather, farmers may either curse or pray for rain—depending on whether or not they have hay down. Our children play in the cool recesses of the grove, squandering this time as if there were an unlimited supply of warm, lazy afternoons.

Fall arrives with a curl of tangy wood smoke rising from a chimney. The crops have ripened, turning the countryside into a patchwork of russet and gold.

Here is the season of harvest, a time for gathering in against the future. September brings the autumnal equinox; the days swiftly grow shorter. The trees have donned their finery, splashes of ruby and amber against the sapphire sky. The evening air turns crisp, and sound carries farther. A freight train laden with fall's bounty blows its mighty air horn. I can hear the lonesome moan echoing across the empty miles.

My neighbor harvests soybeans in the gathering dusk, his gargantuan combine belching a thunderhead of dust that hangs in the still air. There's a whistle of wings and I glance up to see a flock of teal streak over. I watch until they become tiny specks on the southern horizon. Winter sometimes sneaks in on cat's paws, dusting the countryside with bone-colored powder.

The season can also scream in like an ice-blooded Fury, with swirling winds that slash through even my thickest clothing.

The sun becomes a snowbird, lingering for most of the day in the southern part of the sky. My only company during morning and evening chores are the stars—sentinels who look down upon me, cold and unblinking, across the light years. But winter is also a time for celebrations, of family gatherings and sumptuous food and jovial company. Nothing is more delightful than coming in from the deep cold and smacking into a wall of steamy, luscious cooking aromas. This is one of civilization's finest achievements.

The rhythms of life slow. Winter is a season for rest and for early bedtimes.

Each night, an airplane wings its way over our farm on its scheduled voyage to somewhere. Sometimes I'll lie awake beside my slumbering wife and await its arrival. I can hear the Doppler shift in its tone as it drones on by. I think about how lonely it must be up there in the cockpit, to be awake while others sleep, to tunnel through the infinite blackness of the winter night. I wonder if the pilot ever thinks about those below. I push these thoughts aside, snuggle up to my wife, and pull the covers closer.

And earth and I both find rest and pass the long winter night dreaming of spring. ●

Acknowledgments

~~~~~~~~~~~~~~~~~~~~~~

First and foremost, I want to thank my entire family for their steadfast support throughout the many years it took to produce this work. A special shout-out to our sons, Paul and Chris, for providing me with such vast amounts of writing fodder.

Thank you also to Arielle Eckstut. Arielle and her husband, David Henry Sterry, and their Pitchapalooza program helped me plant the seed that grew into this book.

I would like to express my deep gratitude to Danielle Svetcov, my literary agent at the Levine Greenberg Rostan Literary Agency. This work wouldn't have been possible without Danielle's wisdom and encouragement.

A heartfelt thanks to Bruce Tracy, my editor at Workman Publishing, for his gentle and thoughtful guidance. He truly helped make this book wonderfuler.

Thanks to Dan Brown, a dear friend and my former high school English teacher. Dan is a prime example of how the best teachers can continue to influence their students long after class is over.

And finally, I would like to say "Thanks, pardner" to

Mel Kloster, the county agent guy who first advised me to publish my work. Mel is now rounding up strays in that big cow pasture in the sky, but his memory continues to live on.

About the Author

J erry Nelson is a freelance writer and former dairy farmer. His works have been published in the nation's top farm magazines, including *Successful Farming*, *Farm Journal*, *Progressive Farmer*, and *Living the Country Life*. For nearly twenty years, he has penned a weekly newspaper column called Dear County Agent Guy. Jerry's column reaches 250,000 readers each week. In addition to print media, Jerry's column is published on numerous newspaper websites. *Successful Farming* also posts his column on their website, agriculture.com.

Garrison Keillor has used several of Jerry's scripts on the nationally syndicated radio program *A Prairie Home Companion*. Jerry has been featured on South Dakota Public Radio and Television.

After leaving the dairy farming business in 2002, Jerry took a position as a writer/ad salesman for the *Dairy Star*, a bimonthly newspaper for dairy operators all across the Midwest. Jerry and his wife, Julie, live in Volga, South Dakota, on the farm that Jerry's great-grandfather homesteaded in the 1880s.